Abhinand Jha

Soil erosion quantification using cosmogenic radionuclide Beryllium-7

Abhinand Jha

Soil erosion quantification using cosmogenic radionuclide Beryllium-7

Südwestdeutscher Verlag für Hochschulschriften

Impressum / Imprint
Bibliografische Information der Deutschen Nationalbibliothek: Die Deutsche Nationalbibliothek verzeichnet diese Publikation in der Deutschen Nationalbibliografie; detaillierte bibliografische Daten sind im Internet über http://dnb.d-nb.de abrufbar.
Alle in diesem Buch genannten Marken und Produktnamen unterliegen warenzeichen-, marken- oder patentrechtlichem Schutz bzw. sind Warenzeichen oder eingetragene Warenzeichen der jeweiligen Inhaber. Die Wiedergabe von Marken, Produktnamen, Gebrauchsnamen, Handelsnamen, Warenbezeichnungen u.s.w. in diesem Werk berechtigt auch ohne besondere Kennzeichnung nicht zu der Annahme, dass solche Namen im Sinne der Warenzeichen- und Markenschutzgesetzgebung als frei zu betrachten wären und daher von jedermann benutzt werden dürften.

Bibliographic information published by the Deutsche Nationalbibliothek: The Deutsche Nationalbibliothek lists this publication in the Deutsche Nationalbibliografie; detailed bibliographic data are available in the Internet at http://dnb.d-nb.de.
Any brand names and product names mentioned in this book are subject to trademark, brand or patent protection and are trademarks or registered trademarks of their respective holders. The use of brand names, product names, common names, trade names, product descriptions etc. even without a particular marking in this works is in no way to be construed to mean that such names may be regarded as unrestricted in respect of trademark and brand protection legislation and could thus be used by anyone.

Coverbild / Cover image: www.ingimage.com

Verlag / Publisher:
Südwestdeutscher Verlag für Hochschulschriften
ist ein Imprint der / is a trademark of
OmniScriptum GmbH & Co. KG
Heinrich-Böcking-Str. 6-8, 66121 Saarbrücken, Deutschland / Germany
Email: info@svh-verlag.de

Herstellung: siehe letzte Seite /
Printed at: see last page
ISBN: 978-3-8381-3958-6

Zugl. / Approved by: Bremen, Universitatet, Diss., 2013

Copyright © 2014 OmniScriptum GmbH & Co. KG
Alle Rechte vorbehalten. / All rights reserved. Saarbrücken 2014

Contents

List of Tables	4
List of Figures	6
ABSTRACT	13
Thesis Outline	17

1 GENERAL INTRODUCTION AND PHYSICAL BACKGROUND

1.1.	Motivation	19
1.2.	Soil erosion processes by water	21
1.3.	Environmental tracers in soil erosion studies	24
1.3.1.	Vertical migration of radionuclides in soils	25
1.3.2.	Diffusion of radionuclides in soils	26
1.4.	Cosmogenic ^{7}Be in the environment	29
1.4.1.	Nuclear properties	30
1.4.2.	Activity measurement	30
1.4.3.	Production and delivery to earth's surface	31
1.4.4.	Aerosol size distribution	32
1.4.5.	Atmospheric residence time and concentration	32
1.4.6.	Stratosphere-troposphere exchange and seasonal variability	33
1.4.7.	Atmospheric deposition	33
1.4.8.	Distribution in freshwaters	35
1.4.9.	Distribution in vegetation and soils	35
1.5.	^{7}Be: a promising short-term soil erosion/deposition tracer	38
1.5.1	Literature survey and state of the art	38
1.5.2	Key considerations of current erosion/deposition estimation technique using ^{7}Be and its limitations	39
1.5.3.	Scope of further development of the ^{7}Be technique for soil erosion/ deposition quantification	41

1.6.	Research objectives	41

2 VERTICAL TRANSPORT MODEL FOR ^{7}BE IN SOIL

2.1.	Introduction	44
2.2.	One dimensional diffusion model	45
2.2.1.	General assumptions of the model	45
2.2.2.	Model conceptualization	45
2.2.3.	1-D diffusion equation with flux boundary condition (4a)	46
2.2.4.	1-D diffusion equation for Pulse boundary condition (4b)	48
2.3	Total inventory of ^{7}Be at the reference sites	49
2.3.1	Time evolution of the total inventory and depth distribution of ^{7}Be at the eroded point	50
2.4	Estimation of short-term erosion rates using ^{7}Be diffusion model	54
2.4.1	Steady state approach	54
2.4.2	Non steady state approach	55
2.4.2.1.	Crank Nicolson scheme	55
2.4.2.2.	Erosion quantification using Crank-Nicolson scheme	57

3 MATHEMATICAL SIMULATIONS OF THE SYSTEM UNDER STUDY WITH ^{7}BE DIFFUSION MODEL

3.1.	Time evolution of ^{7}Be total inventory	59
3.2.	Time evolution of the depth distribution of ^{7}Be	61
3.3.	Time evolution of the ^{7}Be total inventory and depth distribution for varying input fluxes J	62
3.4.	Time evolution of the ^{7}Be depth distribution after ploughing	65
3.5.	^{7}Be distribution in soil for a pulse-like input	66
3.6.	Summary	66

4 FIELDWORK AND LABORATORY ANALYSIS

4.1.	Introduction	68
4.2.	Study site Müncheberg	68
4.3.	Climate and soil characteristics at Müncheberg	71

4.4.	Sampling design	72
4.5.	^{7}Be measurement program for erosion/deposition quantification	73
4.6.	Sampling methodology	74
4.6.1.	Total inventory sampling using Gauge-Scraper plate method	75
4.6.2.	Fine increment soil sampler for depth distribution measurements	78
4.7	Sample processing	80
4.7.1.	Air drying, grinding, annihilation and sieving for total inventory	80
4.7.2.	Fine depth soil sample processing	81
4.8	Sample analysis	81
4.9	Summary	85

5 RESULTS AND DISCUSSION

5.1.	Introduction	86
5.2.	^{7}Be at the reference sites at Müncheberg	86
5.2.1.	Depth distributions of ^{7}Be at the study site	86
5.2.2.	Grain size characteristics of suspended sediments	92
5.2.3.	Soil inventories of ^{7}Be at the reference sites	93
5.3.	^{7}Be measurements for soil erosion assessment at the study plot	98
5.3.1.	Temporal changes of ^{7}Be activities at the study plot	99
5.3.2.	^{7}Be measurements in the soil at tin barrier and V-channel	101
5.3.3.	Estimation of soil redistribution at the study plot	105
5.3.3.1.	Erosion rates at the tilled plots	109
5.3.3.2.	Erosion rates at the no-till plots	114
5.3.4.	Comparison of erosion rates calculated by ^{7}Be and direct soil measurements from ZALF	118

6 CONCLUSIONS AND OUTLOOK

6.1	Conclusions	123
6.2	Outlook	124
APPENDIX		**126**
BIBLIOGRAPHY		**145**

List of Tables

Table 1:	Radionuclides used for studying soil erosion/deposition rates	24
Table 2:	Spallation reactions of protons and neutrons	30
Table 3:	Annual atmospheric deposition of ^{7}Be and rainfall at different locations	34
Table 4:	Terrestrial inventories of ^{7}Be	37
Table 5:	Total inventory evolution of ^{7}Be at an eroding position	51
Table 6:	Values of parameters taken for simulations	59
Table 7:	Physical soil properties at the study site	71
Table 8:	Sampling chart for soil samples at the site ZALF, Müncheberg	76
Table 9:	Description of the study site and reference sites	77
Table 10:	Germanium detectors used for the gamma spectroscopic measurements	82
Table 11:	Values of fit parameters for ^{7}Be at reference site 1 in Müncheberg	91
Table 12:	Statistical summary for parameters D and h_0	91
Table 13:	Estimated diffusion coefficients and penetration depths for different soils	92
Table 14:	Total inventories of ^{7}Be for different soil conditions	96
Table 15:	Monthly reference inventories (A_{REF}) of ^{7}Be at Müncheberg	97
Table 16:	Erosion events and plot characteristics at Müncheberg	98
Table 17:	Soil redistribution documented for the study site for 12 rainfall events during the years 2010 and 2011, based on ^{7}Be measurements at the tilled plot. The soil redistribution estimates are compared with the physical soil measurements at the study plot.	106
Table 18:	Soil redistribution based on ^{7}Be measurements for the tilled plot at the study site for heavy rainfall events in 2010-11.	107

Table 19: Soil redistribution documented for the study site for 12 rainfall events during the years 2010 and 2011, based on ^{7}Be measurements at no-till plot. The soil redistribution estimates compared with the physical soil measurements at the no-till plot 117

Table 20: Soil redistribution based on ^{7}Be measurements for the no-till plot at the study site for heavy rainfall events in 2010-11. 118

Table 21: Soil budget and comparison of estimated soil using the ^{7}Be technique and direct measurements at the tilled plot. 120

Table 22: Soil budget and comparison of estimated soil using the ^{7}Be technique and direct measurements at the no-till plot. 121

List of Figures

Figure 1:	Water erosion vulnerability map (USDA, 2008) [7]	20
Figure 2:	Top: Different types of erosion processes on an exposed slope [14]. Bottom: Erosion features and development of rills [110]; A: Surface erosion with arrows indicating direction of the flow, B: Prerills, C, D: Incision rills, E: Channel rills.	22
Figure 3:	(1) Penetration profile of ^{137}Cs (2) Penetration profile of ^{7}Be ((1) and (2) are constructed from the data collected by the author)	26
Figure 4:	Concentration-distance curves for an instantaneous plane source for different times t with $t_1 < t_2$.	27
Figure 5:	Production and deposition of ^{7}Be in the atmosphere	29
Figure 6:	Be-7 concentrations and the sunspot numbers at the period 1987-2003 [58].	31
Figure 7:	Depth distribution of ^{7}Be at different study sites. Figures 1, 2, 3 and 4 are modified by the author from the sources in the literature [30], [33],[19], [94] respectively.	36
Figure 8:	^{7}Be concentration distribution for flux boundary condition in a semi infinite column of soil. The numbers on the curves indicate the half- lives of ^{7}Be.	47
Figure 9:	Depth distribution of ^{7}Be for different migration times. 1, 2, 3, 4, 5, 6 on the curves indicate the migration times 0, 5 days, 10 days, 20 days, 30 days and 40 days respectively.	49
Figure 10:	Hypothetical depth distribution curve for ^{7}Be concentration in soil. Green shaded region indicates the eroded layer of soil Δz. Red curve is the depth distribution of ^{7}Be after an erosion event.	38
Figure 11:	Time evolution of ^{7}Be concentration with depth for four

migration times. Blue curve: ^{7}Be distribution left in the soil after and erosion event. 1, 2, 3 and 4 indicate four migration times in days t = 20, 40, 80, 150 respectively. ... 53

Figure 12: A: Truncation errors for Crank-Nicolson scheme as a function of Δx for fixed Δt. B: Truncation errors for Crank-Nicolson scheme as a function of Δt for fixed Δx. ... 56

Figure 13: Scheme for the calculation of erosion rates for multiple erosion events. E1 and E2 are erosion events at time instances t_1 and t_2 respectively. t'_2 is the time instance before erosion event E2 occurs. ... 57

Figure 14: Time evolution of the total inventory of ^{7}Be for eroded depths Δz of 0.5 mm (Blue curve), 1 mm (Green Curve), 1.5 mm (Red curve) 2 mm (Cyan curve) and 2.5 mm (Violet curve). The curves start from an percentage inventory which is left after an erosion event. ... 60

Figure 15: Time evolution of the total inventory of ^{7}Be for multiple erosion events E1,E2, E3, E4 and E5 with eroding depths of 0.5 mm, 1 mm, 1.5 mm, 2mm and 2.5 mm respectively. ... 61

Figure 16: Time evolution of the ^{7}Be depth distribution after an erosion event with an eroded depth Δz = 1.5 mm. Blue curve: Depth distribution after an erosion event at t = 0; Green curve: depth distribution evolved at t = 35 days; Red curve: depth distribution evolved at t = 60 days; Cyan curve: depth distribution evolved at t = 80 days; Black curve: Steady state depth distribution which overlaps with the numerical depth distribution at t = 90 days. ... 62

Figure 17: Time evolution of the total inventory of ^{7}Be for eroded depths Δz =1.5 mm for different input fluxes using Equation (14): 3.54×10^{-5} (Blue curve);4.54×10^{-5} (Green curve); 5.54×10^{-5}

	(Red curve) and 7.54×10⁻⁵ (Cyan curve). The curves start from an percentage inventory which is left after an erosion event.	63
Figure 18:	Time evolution of depth distribution of ^{7}Be for varying input fluxes (A: J= 3.54×10⁻⁵ Bq m⁻² s⁻¹, B: J= 6.54×10⁻⁵ Bq m⁻² s⁻¹). The numbers on the curves indicate the simulation time in days. Blue curve: ^{7}Be depth distribution after erosion; Black curve: steady state depth distribution of ^{7}Be.	64
Figure 19:	Evolution of the depth distribution of ^{7}Be with time after ploughing. The numbers on the curves indicate the simulation times in days. (1: 10 days, 2: 30 days, 3: 60 days, 4: 100 days, 5: 150 days and 6: 200 days)	65
Figure 20:	Depth distribution of ^{7}Be for three migration times estimated with equation (10). The Blue curve is the pulse-like input at the boundary at t=0. The Green curve, Red curve and Cyan curve indicates the depth distribution at t = 5 days, t=10 days and 15 days respectively.	66
Figure 21:	Agricultural practices using different types of machines; A: No-till machines, B: Conventional ploughing machines	69
Figure 22:	A: The experimental plot at Müncheberg research station. B: Schematic of the Müncheberg study plot. The bold arrow in the middle represents the direction of the slope.	70
Figure 23:	The experimental plots with the barrier and funnel system at the bottom; A: barrier at no-till plot, B: barrier at till plot, C: Funnel system with automatic weather station.	71
Figure 24:	The sampling design for ^{7}Be measurements at Müncheberg with RF1, RF2 and RF3 representing the reference sites. The black arrow in the middle of the plot indicates the slope direction.	72
Figure 25:	Rainfall events with soil sampling. Red dotted stems indicate	

	the sampling events in 2010-11.	74
Figure 26:	A: Gauge and scraper plate at the study site. B: The gauge is placed on the soil and with the help of a scraper plate the soil of a particular depth is cut.	75
Figure 27:	A: Reference site RF3 at Müncheberg. B:Grass covered reference site, RF3	77
Figure 28:	Fine increment cylindrical soil sampler (Photo: Klaus Schmidt)	78
Figure 29:	Sampling technique for the depth distribution of ^{7}Be. A: Material needed for sampling, B: Measuring cylinder hammered into the soil, C: collection of the sample with soil surface on the top, D: cut slices are put into small vessels for drying.	79
Figure 30:	Sample processing for the total inventory of ^{7}Be. A: Aluminium tray used for sample drying. B: Sample spread on the aluminium tray. C: Oven used for drying the sample at 105°C. D: Sample geometry used for measurement.	80
Figure 31:	Sample geometry used for the measurement of the ^{7}Be depth distribution at the reference sites.	81
Figure 32:	High purity Germanium detectors used to measure ^{7}Be at Bundesamt für Strahlenschutz, Berlin, Germany. A: Coaxial detector used for aluminium bottle geometry, B: Ge Well detector for test tube geometry.	83
Figure 33:	A: Gamma spectrum of soil; B: ^{7}Be photopeak shown at 477.6 keV	84
Figure 34:	Depth distribution of ^{7}Be in the dry period (May 2011). Horizontal error bar is the statistical uncertainty of the measurement.	87
Figure 35:	Be-7 concentrations (kBq m^{-3}) with depth for wet months A:	

	June, B: September, C: November and D: October. Blue curve is the statistical fit of the diffusion model (Equation (8)) to the measured data. The dotted curves are the confidence intervals of the fit. Horizontal error bars are the statistical uncertainties of the measurement and the vertical error bars represent the uncertainties over depth.	88
Figure 36:	Cumulative depth distribution of ^{7}Be at reference site in Müncheberg for the months June (A), September (B), November (C) and October (D).	90
Figure 37:	Particle size distribution of suspended sediments discharged during a rainfall event in June 2011 for (A) V-channel and (B) Barrier.	93
Figure 38:	Time series of total inventories of ^{7}Be at the reference sites and Rainfall data for 24 months. In the top plot each group of bar graphs with the measurement uncertainties represents the 3 reference inventories measured in each month with pink, violet and green colours for Reference sites 1, 2 and 3 respectively. Solid red curve is the sinusoidal fit to the data and the dotted lines indicate the 95% confidence bounds to the fit.	94
Figure 39:	Empirical cumulative distribution plots for the ^{7}Be inventories. Red curve: data set for the year 2010 and Blue curve: data set for the year 2011.	95
Figure 40:	Daily precipitation events for the study period from March 2010-September 2011. Violet arrows indicate the erosion events measured at the study plot.	99
Figure 41:	Twelve erosion events occurred during the study period from 2010-2011. Δt represents the time between the erosion events. Colours indicate the erosion processes. (Red: Rill/	

	interrill erosion, Yellow: Surface erosion, Green: Splash erosion)	100
Figure 42:	^{7}Be activities in soils collected in tin barriers and V-channels in 2010	102
Figure 43:	^{7}Be activities in soils collected in tin barriers and V-channels in 2011	102
Figure 44:	^{7}Be activity of the suspended sediment (barrier + V-channel) samples eroded from tilled plots for rainfall events in A: 2010, B:2011. The uncertainties indicate the gamma spectrometry measurement precision at 95% level of confidence.	103
Figure 45:	Erosion processes observed at tilled and no-till plots. A: Rill erosion at tilled plot; B: Surface erosion at no-till plot	104
Figure 46:	Soil collected at the tin barriers and measured ^{7}Be activities at no-till plot for all erosion events. Top: Mass of soil collected at the tin barrier and Bottom: ^{7}Be activities in sediments.	105
Figure 47:	Erosion/deposition rates estimated for 12 erosion events at tilled plot for measurement points A: 35m; B: 50m. ('-': erosion rates, '+': deposition rates)	109
Figure 48:	Time evolution of the total inventory of ^{7}Be for erosion events occurring in A: 2010; B: 2011, at 50 m along the slope length. The numbers between the erosion events represent the time in days between the events. The curves start from an percentage inventory which is left after an erosion event.	113
Figure 49:	Soil redistribution rates estimated for the tilled plot for the rainfall event in May 2010(Top) and Aug-1 2011(Bottom). ('-': erosion rates, '+': deposition rates)	114

Figure 50: Erosion/deposition rates estimated at no-till plot for 12 erosion events for measurement points 35m (Top plot) and 50m (bottom plot). ('-': erosion rates, '+': deposition rates)   115

Figure 51: Soil budget at the study plot.   119

ABSTRACT

Die Fallout-Radionuklide, wie z. B. ^{137}Cs, ^{210}Pb$_{ex}$, sind weitverbreitet zur Quantifizierung von Bodenerosions- und Sedimentverteilungsraten innerhalb landwirtschaftlicher Nutzflächen bzw. Gewässern. Die mit diesen Radionukliden ermittelte räumliche und zeitliche Bodenverlagerung gilt als eine wertvolle Ergänzung zu konventionellen Methoden zur Bestimmung der Bodenerosion. Mit der Auswertung der oben genannten Radionuklide können allerdings nur mittelfristige (40 bis 100 Jahre) Bodenerosionsraten bestimmt werden. Die ^{7}Be Methode besitzt das Potenzial kurzzeitige meteorologische Ereignisse (z. B. Starkregen) mit sich daraus ergebenen Erosionsraten in Verbindung zu setzen und diese abzuschätzen. Die ^{7}Be Methode gewinnt in einer Zeit des sich abzeichnenden Klimawandels, der Veränderung der Landnutzung und andere menschliche Aktivitäten zunehmend an Bedeutung.

Die vorliegende Arbeit stellt ein mathematisches Modell vor, das auf den physikalischen Prozessen der molekularen Diffusion, unter Berücksichtigung des radioaktiven Zerfalls des ^{7}Be, basiert, um die vertikale Bewegung von ^{7}Be in Böden zu untersuchen. Mit diesem Modell wurden Erosionsraten für 12 einzelne Niederschlagsereignisse über einen Zeitraum von zwei Jahren in dem Untersuchungsgebiet Müncheberg, Deutschland quantifiziert.

Das Modell geht von der Annahme aus, das es eine pulsähnliche Fallout-Anfangsbedingung gibt, bei der vor Beginn der Betrachtung keine ^{7}Be-Aktivitätskonzentration vorlag. Nach dem Niederschlagsereignis wird mit Hilfe einer angenommenen Exponentialverteilung der ^{7}Be-Aktivitätskonzentration auf der untersuchten landwirtschaftlichen Fläche der Diffusionskoeffizient D ermittelt. Dabei wurde das Modell mit mehr als 15 Tiefenverteilungen ausgestattet, um den effektiven Diffusionskoeffizient D besser abschätzen zu können. Für den Diffusionskoeffizenten D konnten Werte in der Größenordnung von $10^{-12} - 10^{-13}$ m^2 s^{-1} für lehmig bis sandige Bodenarten ermittelt werden. Die Boden-Analysen zeigen, dass der Diffusionskoeffizient D nicht nur ein Anpassungsparameter ist, sondern auch von den physikalisch-chemischen Eigenschaften der Radionuklide im Boden abhängig ist.

Die Bodenabtragsraten bei der Fläche des „konventionell betriebenen Anbaus" (Pflügen und sonstige Bodenbearbeitung) im Untersuchungsgebiet lagen zwischen kleiner 0,001 bis 4,7 ± 0,4 kg m^{-2} und bei der Fläche der „Direktsaat" (keine Bodenbearbeitung) zwischen 0,30 ± 0,05 kg m^{-2} bis 2,0 ± 1,4 kg m^{-2}. Die abgeschätzte Erosionsrate auf der Fläche der Direktsaat betrug weniger als die Hälfte bezogen auf die Fläche des „konventionellen Anbaus".

- Das entwickelte mathematische Modell in dieser Studie beschreibt den Transport von ^{7}Be in Böden. Es ist das erste umfassend vorgeschlagene Modell, das trotz der vielen Vereinfachungen, z. B. durch die Annahme einer exponentiellen Verteilung des ^{7}Be innerhalb der Profile, die Bodenerosion in gestörten Bodenoberflächen (konventioneller Anbau) sowie ungestörten Bodenoberflächen (Direktsaat) und auf Referenzflächen korrekt wieder gibt.

- Es wurde nachgewiesen, dass der wichtigste physikalische Prozess, der ^{7}Be im Boden transportiert, die molekulare Diffusion ist. Der Nachweis musste unter Berücksichtigung des radioaktiven Zerfalls des ^{7}Be mit seiner vergleichsweise kurzen Halbwertszeit von 53,23 Tagen durchgeführt werden. Migrationsparameter und Messungen bestätigen, dass Sorption der wichtigste physikalische Prozess ist, der die ^{7}Be-Konzentration zu der Bodenoberfläche abgrenzt.

- Das aktuell vorgeschlagene Modell unter Verwendung von ^{7}Be als Tracer wurde erfolgreich nach einzelnen Niederschlägen getestet und konnte an Hand nachfolgender Niederschlagsereignisse weiter modifiziert werden.

- Unter zur Hilfenahme der ^{7}Be Methode konnte erfolgreich zwischen der Rill-Interrill-, Splash und Oberflächenerosion auf der Versuchsfläche unterschieden werden. Zu beachten bleibt, dass das diskutierte Diffusions-Modell in dieser Studie Vegetation auf den Anbauflächen nicht berücksichtigt. In einigen Fällen führt dies bei der Abschätzung der Bodenabtragsrate zu einer Überbewertung.

ABSTRACT

The fallout radionuclides ^{137}Cs, ^{210}Pb$_{ex}$ are used widely for obtaining quantitative information on soil erosion and sediment redistribution rates within agricultural landscapes, over several spatial and temporal scales, and they are frequently seen to represent a valuable complement to conventional soil erosion measurement techniques. However, measurements of these radionuclides provide estimates of medium term (i.e. 40-100 years) soil erosion rates. The shorter-term perspective provided by the ^{7}Be method has the potential to estimate soil erosion rates associated with individual events or short periods. The ^{7}Be method has become increasingly relevant in an environment impacted by climate change, changing land use and other human activities.

The present work establishes a mathematical model based on the physical processes of molecular diffusion and radioactive decay, to study the vertical behaviour of ^{7}Be in soils. This model was further used to quantify erosion rates for 12 individual erosional events over a period of two years at our study site in Müncheberg, Germany.

The scope of the model was explored analytically as well as numerically for Pulse-like fallout initial condition, zero concentration initial condition and exponential distribution initial condition. The model was fitted to more than 15 depth distributions and the resulting model parameter, effective diffusion coefficient D, is evaluated. In general diffusion coefficients estimated were of the order of $10^{-12} - 10^{-13}$ m^2 s^{-1} for loamy to sandy soil types. Diffusion coefficients estimated for our study site were about 10^{-13} m^2 s^{-1}. The soil analyses indicate that the diffusion coefficient D is not merely a fitting parameter, but is related to the physico-chemical properties of radionuclide transport in soils.

The erosion rates estimated at tilled and no-till plots at our study site were between < 0.001 - 4.7 ± 0.4 kg m^{-2} and 0.3 ± 0.5 kg m^{-2} - 2.0 ± 1.4 kg m^{-2} respectively. The magnitude of erosion rates estimated at the no-till plots was less than that at the tilled plots. The main conclusions of this work are:

- The mathematical model developed during this study describes the transport of ^7Be in soils. It is the first extensive model proposed so far that despite of its many simplifications, adequately represents the exponential distribution of ^7Be profiles at disturbed and undisturbed or reference sites.
- Main physical processes, which transport of ^7Be in soil are, diffusion and radioactive decay. Migration parameters and measurements confirm that sorption is the main physical process, which confines ^7Be concentration to soil surface.
- Current erosion estimation methods with ^7Be available in the literature for estimating erosion rates for single rainfall event was successfully modified to quantify erosion rates for multiple rainfall events.
- Erosion rates estimated with ^7Be technique were successfully used to differentiate between the rill-interrill, splash and surface erosion at the study plot.
- The Diffusion model proposed in this study does not take into account the vegetation cover and thus overestimates the erosion rates or in some cases shows the occurrence of deposition on the plot.

THESIS OUTLINE

This PhD thesis is organized into six chapters followed by an APPENDIX section. The chapters are organized in such a manner that they explain the building blocks of the thesis one by one but at the same time are well connected with each other maintaining the flow of the research topic.

Chapter 1 gives a brief background of the soil erosion problem in the world and introduces the basic concepts in the use of radionuclides in soil erosion quantification. A detailed discussion on ^7Be in the environment is done in this chapter. The state of the art of erosion quantification technique using ^7Be is also given here. The research goals of this study have been addressed at the end of this chapter.

Chapter 2 introduces the vertical transport model of ^7Be in soils. The model is tested for different initial and boundary conditions and the analytical solutions to the differential equations are presented here. The non steady-state approach of erosion quantification is explored in this chapter with the use of Crank-Nicolson scheme of numerical discretization of heat equation.

Chapter 3 deals with the mathematical simulations of the ^7Be diffusion model. The model is exploited with the help of different input parameters and the simulations results are presented in the systematic order.

Chapter 4 outlines the study area and the sampling strategy used for this research. A detailed discussion on the sampling methods for total inventory and depth distribution of ^7Be is explained in details. Final section of this chapter focuses on the sample analysis and ^7Be measurement using Gamma Spectroscopy.

Chapter 5 focuses on the detailed analysis of the ^7Be data obtained during this research project. At first, the depth distributions of ^7Be are studied with the help of diffusion model and the estimated diffusion parameters are presented. Secondly, erosion quantification was carried out with the help of diffusion model and ^7Be activities measured at the study site. The estimated short-term soil erosion rates at the study sites for 12 discrete erosion events at tilled and no-till plots are presented in this chapter.

Chapter 6 highlights the main conclusions reached in the thesis and a summary of suggested further research.

Appendix section contains detailed solutions of the differential equations used in this thesis. A separate section is devoted here for the error analysis using Monte-Carlo technique. Computer codes written for the numerical schemes as well as the error analysis are included in this section.

1 GENERAL INTRODUCTION AND PHYSICAL BACKGROUND

1.1 Motivation

Soil erosion by water is one form of soil degradation and has become an important environmental problem. It is beginning to be recognized for being not simply a farming problem but with implications for the wider civil society. More than three quarters of the surface land area affected by erosion is located in the developing countries of Africa, Asia and Latin America, with about one-half of the total occurring in Asia [1] (Figure 1). In the European Union, an estimated 115 million hectares (12% land area) is subjected to water erosion [2]. Unlike other regions of the world, where extensive agriculture is still significant, during the last half century agricultural areas in some parts of Europe e.g. Italy, United Kingdom, Spain have suffered an important decrease. Erosion rates vary a great deal spatially and intense agricultural practices is one of the driving forces that can accelerate erosion.

Frequent cultivation of the plots changes the soil properties and eventually the plots itself. These changes can be positive or negative for soil protection from erosive agents. This depends on the climate, features of the terrain and the canopy cover. In the Spanish mountains, farmers modified the terrain by constructing terraces and ditches in order to increase yield and protect the valuable top-soil. However land abandonment resulted in the degradation of terraces and ditches [3]. In Europe intensively farmed areas are being shifted to intensively forested areas, because of many abandoned hillslopes an extensive policy of reforestation was encouraged [4].

Current concerns about both on-site and off-site adverse effects associated with accelerated soil loss generate an urgent need for obtaining reliable quantitative data on the extent and actual rates of soil erosion worldwide [5, 6]. Rapid and reliable methods of documenting soil erosion and soil degradation within agricultural areas are necessary for providing information for farmers, local governments and risk assessors for adopting the best management practices for soil and water conservation. It is also necessary to investigate the erosion processes for developing land-use policies with the help of soil erosion/sedimentation prediction models, assessment of the economic and environmental impacts of erosion and to select effective soil conservation measures and land-management strategies.

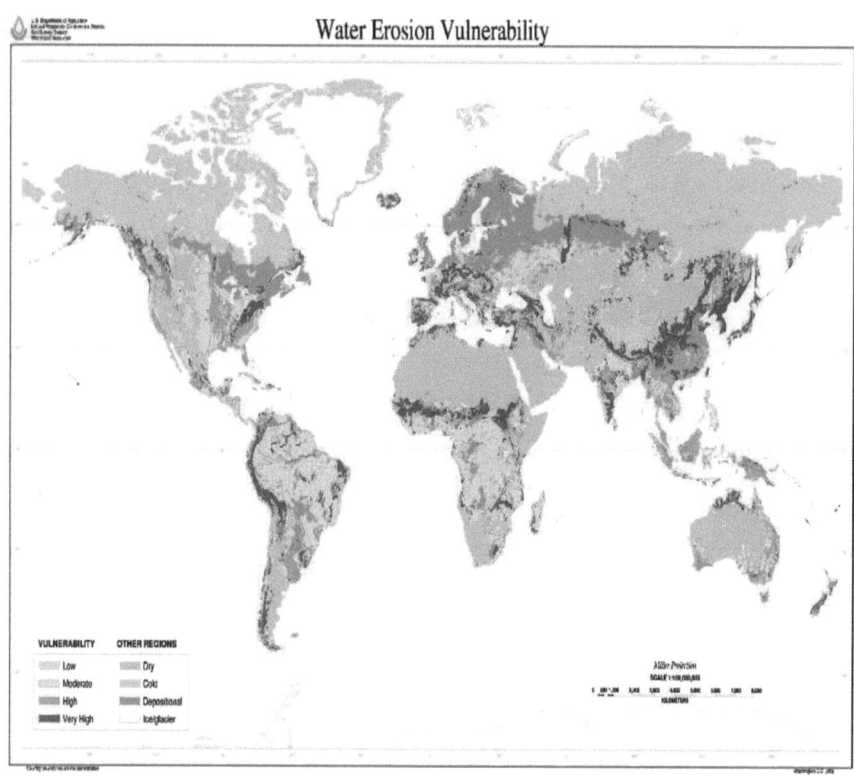

Figure 1: Water erosion vulnerability map (USDA, 2008), [7].

Despite extensive literature on the global, regional and national problems of soil erosion, quantitative and reliable data on the extent and rates of soil erosion are scarce for many regions of the world [8]. Existing methods to assess soil erosion are grouped into two categories: (1) Erosion modelling and prediction methods and (2)

Erosion measurement methods. In both these cases there is a need for direct measurements of soil erosion, which can be done using erosion plots, surveying methods and nuclear techniques. Existing classical techniques such as erosion plots and surveying methods for monitoring soil erosion are capable of meeting some of the requirements, but they have a number of important limitations in terms of representativeness of the data obtained and potential to provide information on long term soil erosion rates and associated spatial patterns over extended areas, and the costs involved [9, 10]. The request for alternative techniques of soil erosion assessment to complement erosion plots has directed the attention to the use of radionuclides such as ^{137}Cs, ^{210}Pb, ^{7}Be and ^{14}C.

The objectives of this chapter are (1) to introduce different erosion processes initiated by water; (2) to describe briefly the use of radionuclides as tracers of soil erosion/deposition; (3) To overview the achievements in the use of cosmogenic ^{7}Be to estimate short-term soil loss for single rainfall events (4) to introduce research goals for the development of ^{7}Be technique to quantify soil erosion/deposition rates during multiple rainfall events.

1.2 Soil erosion processes by water

Soil erosion by water is a complex time-variant process which occurs in three phases, with the detachment of individual particles from soil mass as the first phase followed by their transport by erosive agents such as water as the second phase. When the sufficient energy is not available the third phase of particle deposition occurs [11]. Tillage plays an important source of energy, which leads to the transport of soil downslope [12].

During the early stages of a heavy rainfall event, processes that occur on the field include surface and splash erosion. As the event proceeds, the flow frequently becomes concentrated, and rills are developed (Figure 2 (Top)). Sediment that is detached from the interrill areas moves laterally to the rills in the thin interrill sheet flow [13]. Direct splash to the rills or downslope is not a major mode of transport.

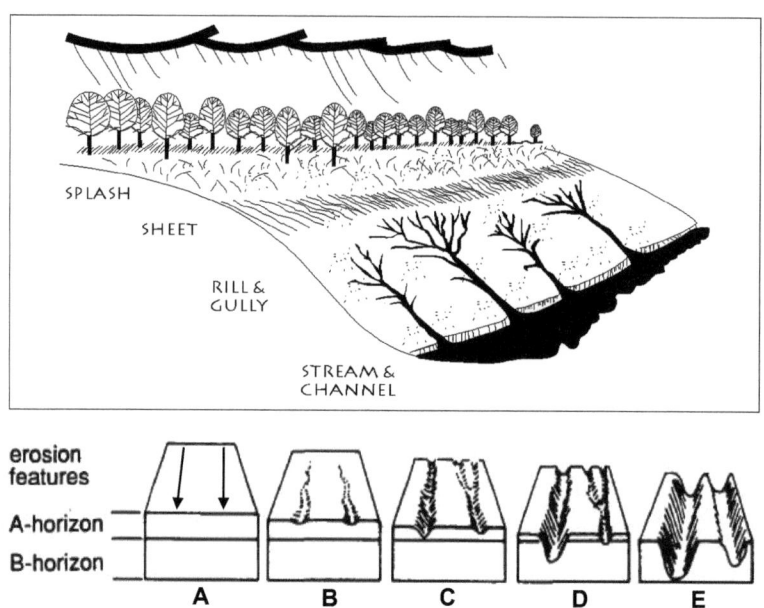

Figure 2: Top: Different types of erosion processes on an exposed slope [14].
Bottom: Erosion features and development of rills [110]; A: Surface erosion with arrows indicating direction of the flow, B: Prerills, C, D: Incision rills, E: Channel rills

Rills as shown in Figure 2 (bottom) are the cracks in the soil, which can reach depths of 1 cm-10 cm [112] and are produced due to natural topographical features, soil roughness, or tillage marks and tracks. Shear and flow velocity are two parameters often used to measure the erosive potential of rill flow. Erosion from areas between the rills is defined as interrill erosion. In Figure 2 (bottom) soil columns are divided into horizons, A and B , depending on the soil characteristics. Horizon A is the top layer and is the zone of major biological activity and is therefore generally enriched with organic matter and typically darker in color than the underlying soil. The horizon B, where some of the materials (e.g. clay, carbonates) that are leached from Horizon A by percolating water tend to accumulate. Horizon B is generally thicker than the horizon A (Figure 2 (bottom)). The clay accumulation and the pressure of overlying soil combine to reduce the porosity of the deeper layers. At our study site the erosion processes given by A, B and C in Figure 2(Bottom) were observed.

Raindrops not only detach soil aggregates and sand (20-2000 µm), silt (20 -6.3 µm), and clay particles (< 2 µm) from the soil mass, but subsequent raindrop impact probably breaks the detached aggregates down further as they are transported to the rills. In addition, raindrops create turbulence within the flow layer, which greatly increases the transport capacity of interrill flow.

Clay is usually considered to be the mineral component of the eroded soil most important in the transport of adsorbed radioactive tracers [15]. The source of the sediment can have a large effect on its chemical composition as well as on the material eroded from the land surface by runoff and delivered to a stream system [16]. The estimation of sediment and associated tracer transport requires information on the size and composition of particles [17]. The grain size composition of the mobilized sediment and the depth within the soil horizons from which it is mobilized result in contrast in the radionuclides and nutrient contents.

In many erosion studies that determined aggregate and primary particle distribution, no differentiation was made between the particle sizes being eroded from rill and interrill areas. Based on the results from the field plots, it was suggested that the particles eroding from interrill areas would generally be smaller than those eroding from the rill areas [15]. Particle selectivity during the erosion process is almost impossible when rill erosion is significant because of the massive removal of particles from the rills [15].

Several laboratory studies using disturbed soils have determined the particle size distribution of interrill sediment [11, 15, 17]. It was found that the interrill sediment was enriched in sand and not in clay, while rill sediment was enriched in clay and not in sand [15]. These findings conflict with the conclusions from other study [13], which reasoned that the particles eroding from interrill areas would generally smaller than those eroding from the rill areas. The sand enrichment in the interrill sediment is attributed to the downward movement of fines in the soil matrix [15]. Until today less information is available on rill erosion and the particle sizes, which can be transported. It was concluded that about 15% of the particles transported in rill flow from a tilled soil (6% slope) was larger than 1 mm [13]. Almost 3% of the sediment was larger than 5 mm, which indicates that rill flow can transport very large particles. Selective erosion under these conditions is highly unlikely. Until today the information on the sizes of particles detached and transported by rill and interrill erosion processes is not complete and is somewhat contradictory.

1.3 Environmental tracers in soil erosion studies

In soil erosion research, both extent and source of soil loss can be determined by artificially labelling the soil particles with appropriate radioactive tracer. Several artificial radionuclides, mainly gamma-ray emitters such as ^{59}Fe, ^{46}Sc, ^{110}Ag, ^{198}Au, ^{134}Cs, ^{51}Cr etc., have been used as tracers in field erosion studies. The majority of radionuclide applications are related to environmental radionuclides such as ^{137}Cs, ^{210}Pb and ^{7}Be, which showed a great potential in assessing soil erosion and deposition [1, 18, 19, 20, 21, 22, 23, 24, 25]. Some of these radionuclides are produced by cosmic rays in the atmosphere (^{7}Be, ^{14}C, ^{32}Si, ^{26}Al and ^{36}Cl), others are members of the natural decay series of the primordial radionuclides ^{238}U, ^{235}U and ^{232}Th (e.g. ^{210}Pb). Artificial radionuclides such as ^{134}Cs, ^{137}Cs that have been released into the environment by nuclear weapon tests and nuclear facilities have also proved very useful [5]. The usefulness of a particular radionuclide depends on (i) its half-life and (ii) the sustainability of the model used for the evaluation of the measured radionuclide values.

Table 1: Radionuclides used for studying soil erosion/deposition rates

Radionuclide	Half-life	Radionuclide Origin	Erosion assessment
^{137}Cs	30.2 years	Man-made	Medium term (~ 40 y)
^{210}Pb	22.3 years	Natural geogenic	Long term (~100 y)
^{7}Be	53.3 days	Natural cosmogenic	Short term (< 30 days)

When radionuclides such as ^{137}Cs, ^{210}Pb and ^{7}Be reach the soil surface through wet and dry deposition, they are quickly and strongly adsorbed by exchange sites of soil particles and become essentially non-exchangeable in most environments. After deposition these radionuclides migrate into the soil column by a number of physical, chemical and biological processes. Accurately measuring these radionuclides in soil/sediment samples is relatively easy using modern instrumentation (high-purity germanium gamma spectrometry) and standardized protocols for quality assurance control. The different time-scales for which the radionuclide technique is applicable for erosion studies are summarized in the Table 1.

Out of the radionuclides demonstrated in Table 1, ^{137}Cs have been successfully used to quantify erosion and deposition processes since the 1970s [6, 25, 26, 27]. About 4000 research papers dealing with the use of ^{137}Cs for soil erosion/deposition rates estimation were published showing that it is a valuable complement to conventional erosion measurement techniques [28].

In the environment impacted by changing climate, changing land use and other human activities it becomes increasingly relevant to document short-term erosion/deposition rates. Thus there is a need of documenting soil redistribution occurring within individual events or short periods. Cosmogenic ^{7}Be offers an advantage of providing estimates of short-term soil erosion/deposition rates. The rest of the thesis will focus on Beryllium-7 and its use in calculating short-term erosion/deposition rates.

1.3.1 Vertical migration of radionuclides in soils

Various radionuclides are incorporated differently into the soil profile (Figure 3) depending upon their respective half-life and history of fallout as well as the history of land-use [29].

In soils ^{7}Be is concentrated near the surface (0–2 cm) as it is adsorbed onto the clay minerals after it is deposited on the soil surface and does not penetrate deeper into soils due to its short half-life. ^{137}Cs values have the greatest activity at depth below surface, reflecting peak delivery of it in the 1960s and/or 1986 due to the Chernobyl accident. Because ⁷Be and ¹³⁷Cs have different distributions in the soil profile, erosion of the soil to different depths will yield characteristic assemblage of radionuclides in the eroded material. Shallow erosion produces proportionally larger amounts of ⁷Be as this radionuclide is concentrated near the surface.

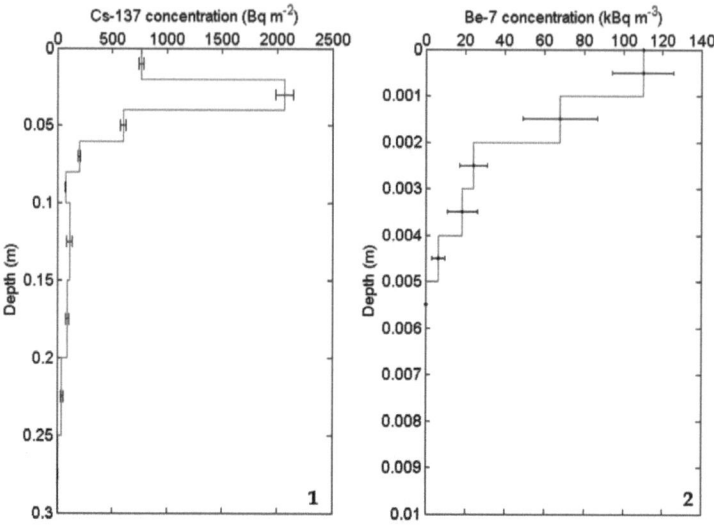

Figure 3: (1) Penetration profile of ¹³⁷Cs (2) Penetration profile of ⁷Be
((1) and (2) are constructed from the data collected by the author)

Deeper incision yields progressively no additional ⁷Be below about 1 cm. The distinct distributions of radionuclides permit, in principle, the use of multiple mass balances to quantitatively estimate the amount of rill and surface erosion and the characteristic depth of erosion associated with each mechanism [30, 31].

1.3.2 Diffusion of radionuclides in soils

The basic processes controlling mobility of radionuclides in soil include convective transport by flowing water, dispersion caused by spatial variations of convection velocities, diffusive movement within the fluid, and physio-chemical interaction with the soil matrix. Many field studies have been done to establish the depth distribution of ^{7}Be in soils [30, 32, 33, 34, 35]. The field observations from the current study and from the published data show that maximum mass activity density (Bq kg^{-1}) of ^{7}Be is found at the point of input of the radionuclide i.e. at the surface of the soil column and is decreased as we go deeper into the soil column. This is a typical profile of purely diffusive transport. From here onwards ^{7}Be transport in soil will be discussed on the basis of diffusive flow.

It is well studied that diffusion process is always active, with spreading essentially a function of time, while the convection and dispersion are a direct consequence of fluid flow in the porous medium, with spreading basically a function of travel distance [36].

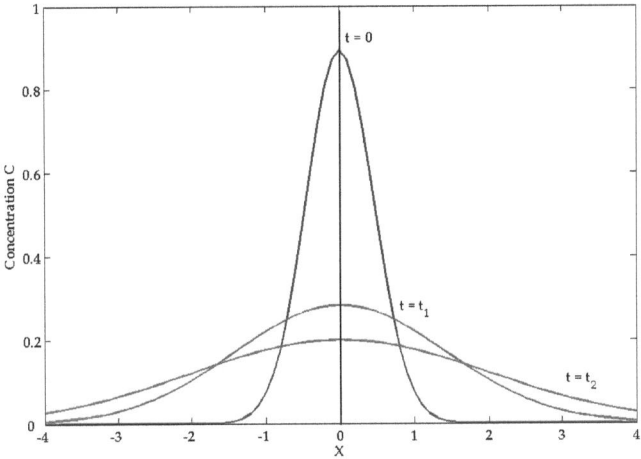

Figure 4: Concentration-distance curves for an instantaneous plane source for different times t with $t_1 < t_2$.

Consider a trace substance C_0 is inserted into a soil column with no ground water flow at x=0 and t=0. It diffuses in both directions away from x = 0. The diffusion equation and relevant solution are given in [37] as:

$$\frac{\partial C}{\partial t} = D \frac{\partial^2 C}{\partial x^2} \qquad (a)$$

$$C(x,t) = \frac{C_0}{2\sqrt{\pi D t}} e^{-\frac{x^2}{4Dt}} \qquad (b)$$

here D: diffusion coefficient, [m²/s] and C : trace substance concentration [kg m⁻³]

A typical diffusion profile is described by the probability density function of a normal distribution, $f_N(x) = \frac{1}{\sigma\sqrt{2\pi}} e^{-\frac{(x-\mu)^2}{2\sigma^2}}$. Mean and variance of the distribution are given by its expectation values as, $E[x(t)] = 0$ and $\sigma[x(t)] = \sqrt{2Dt}$.

The soil from the diffusion point of view represents a non-homogeneous (in granular and chemical composition) multifunctional adsorbing system. This system consists of three phases: solid phase (soil minerals), liquid phase (soil solution) and gaseous phase (soil atmosphere). The most active and important phases of interaction radionuclides with soil are solid and liquid phases. The solution layer adhering to the soil particles surface and other parts of the soil solution differ in their behaviour. The factors affecting the radionuclide movement in soils are: physiochemical behaviour of migrating radionuclide, adsorption capacity, soil moisture, concentration and composition of soil solution (viz. ions Ca^{2+}, Mg^{2+}, Na^+, K^+, H^+, NH_4^+, Cl^-, SO_4^{2-}, NO_3^-), pH of the soil (diffusion in acidic soil is much faster than in the neutral medium), organic substances content and climatic conditions.

Diffusion in porous media is affected in different ways by the geometry of porous structure and by the contaminant interaction with the pore walls. To account for these effects, an effective diffusion coefficient, D_{eff}, is used to describe the contaminant diffusion in porous media [38].

$$D_{eff} = \frac{n_e}{\tau^2} D_0$$

where, n_e: effective porosity (open and interconnected pores)

τ : Tortuosity – is a measure of the effect of the shape of the flow path

D_0: Diffusion coefficient for particular solute in bulk water.

During this thesis the effective diffusion coefficients for diffusing $^7Be^{2+}$ ions in the soils will be estimated will be denoted by D.

1.4 Cosmogenic ^{7}Be in the environment

Beryllium-7 is a short-lived environmental radionuclide of cosmogenic origin. It is produced in the upper atmosphere by spallation of nitrogen and oxygen by cosmic rays (Figure 5). The nuclear reaction produces BeO or Be(OH)$_2$, which diffuses through the atmosphere until it attaches to atmospheric aerosols. Subsequently it is deposited to the earth surface as wet and dry fallout [34, 39, 40], although available measurements suggest that Be-7 is primarily associated with precipitation [20, 41, 42].

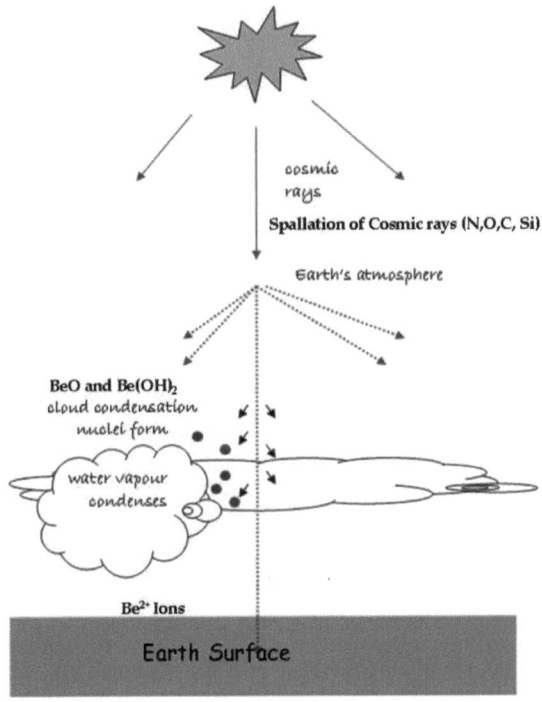

Figure 5: Production and deposition of ^{7}Be in the atmosphere

^{7}Be was first measured in rainwater samples collected at Chicago IL and Lafayette IN (USA) between 1953 and 1954 [43]. The presence of this radionuclide attached to aerosols in surface air was reported soon after [44]. Since these pioneering efforts, numerous studies on the production, concentration, distribution and application of ^{7}Be in the environment have been conducted.

1.4.1 Nuclear Properties

The light atmospheric nuclei, such as carbon, nitrogen and oxygen absorb protons and neutrons of primary components of cosmic rays, according to the reactions in Table 2 [45]. The half-life for the transmutation of $^7Be + e^- \rightarrow\, ^7Li + \nu$ is 53.22 ± 0.06 days [46]. The decay of ^{7}Be can proceed either directly to the ^{7}Li ground state or to the first excited state in ^{7}Li. The branching ratio to the first excited state is 10.44 ± 0.04% [46]. The decay then proceeds to the ^{7}Li ground state by prompt gamma- ray emission with energy of approximately 477.6 keV.

Table 2: Spallation reactions of protons and neutrons

Nuclear reactions	Energetic Particles
1. $^{12}_{6}C + ^{1}_{1}p \rightarrow ^{7}_{4}Be + ^{6}_{2}Li$ 2. $^{14}_{7}N + ^{1}_{1}p \rightarrow ^{7}_{4}Be + 2^{4}_{2}He$ 3. $^{16}_{8}O + ^{1}_{1}p \rightarrow ^{7}_{4}Be + ^{7}_{3}Li + ^{3}_{2}He$	Protons
1. $^{12}_{6}C + ^{1}_{0}n \rightarrow ^{7}_{4}Be + ^{6}_{2}He$ 2. $^{14}_{7}N + ^{1}_{0}n \rightarrow ^{7}_{4}Be + ^{8}_{3}Li$ 3. $^{16}_{8}O + ^{1}_{0}n \rightarrow ^{7}_{4}Be + ^{6}_{2}He + ^{4}_{2}He$	Neutrons

1.4.2 Activity measurement

Berillyum-7 activities are normally determined in environmental samples using gamma spectrometers that detect the 477.6 keV photopeak. In earlier investigations, ^{7}Be was detected using thallium (Tl) activated sodium-iodide (NaI) detectors [43, 47, 48, 49, 50]. However, the relatively low resolution of these instruments requires chemical separation of ^{7}Be before the gamma analysis because a NaI (Tl) detector cannot distinguish the ⁷Be photopeak from other radionuclides with decay in the same energy region (²²⁸Ac at 462 keV; ¹⁰³Ru at 497 keV).

1.4.3 Production and delivery to the earth's surface

Production of ⁷Be is continuous, global in extent and depends on the cosmic-ray flux, which varies with latitude, altitude, and solar activity (Figure 6) [51, 52, 55, 56, 57]. The main production region is the stratosphere (90%), though it is also produced in some abundance within the troposphere (approx. 10%).

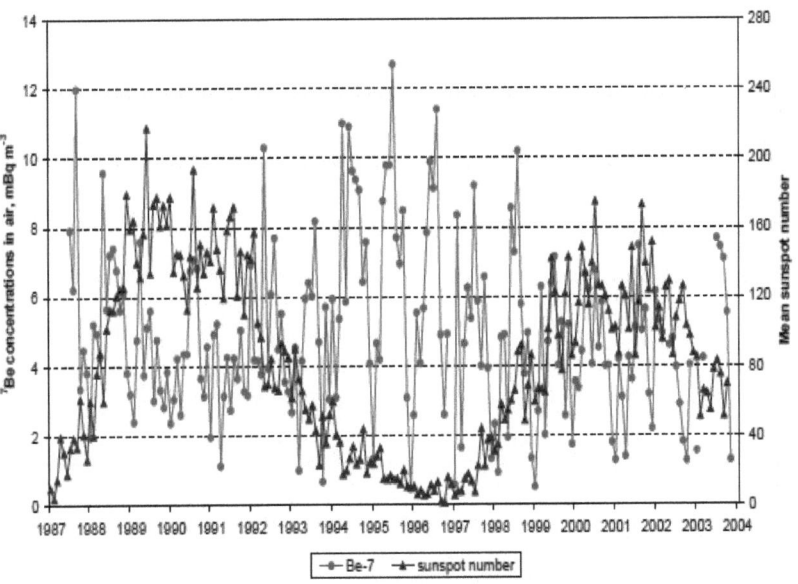

Figure 6: Be-7 concentrations and the sunspot numbers at the period 1987-2003 [58].

Cosmogenic ⁷Be production varies with the 11 - year solar cycle. Solar activity maxima result in increased deflection of cosmic rays from the solar system [52] that decreases the cosmic ray flux to earth, and thus decreases ⁷Be production. Several studies have demonstrated an inverse relationship between cosmogenic ⁷Be concentrations in the air and on earth's surface with solar activity [53, 54]. The ⁷Be production rate in the atmosphere from available data on cosmic-ray-produced neutron and proton fluxes and spallation reactions involving nitrogen and oxygen has a global average value of 810 atoms m^{-3} s^{-1} [52, 55]. This value has been adopted by the United Nations Scientific Committee on the Effects of Atomic Radiation (UNSCEAR) for the purposes of reporting ^7Be production rates in the atmosphere [59]. UNSCEAR reports the annual effective dose due to cosmogenic ^7Be to be 0.03 µSv [59]. This represents around 0.001% of the total annual effective dose due to all natural radiation sources. This shows that ^7Be is a not a health risk to the population.

1.4.4 Aerosol size distribution

Once ^7Be is formed in the atmosphere, it rapidly associates primarily with submicron- sized aerosol particles [60, 61, 62]. ^7Be in these fine aerosols may subsequently enter the marine as well as the terrestrial environment and vegetation via wet or dry depositional events [43]. Following deposition, ^7Be tends to associate with the particulate matter. The ^7Be bearing aerosol is generated through the process of attachment when ^7Be in the form of BeO and BeOH attaches electrostatically to atmospheric dust particles [43, 52]. Transformation of ^7Be-bearing aerosol as it traverses the atmosphere can occur by a number of physical, chemical and meteorological processes (e.g. fog and cloud droplet formation, washout, rainout, sedimentation etc.), which determine the overall activity size distribution of ^7Be on the surface air aerosol population. Measurements indicate that ^7Be-bearing aerosols in surface air follow a unimodal lognormal size distribution and have an activity median aerodynamic diameter between 0.07-2 µm [45, 61, 62, 63].

1.4.5 Atmospheric residence time and concentration

The stratospheric residence time of aerosols is approximately 14 months [64]. This exceeds the half-life of ^7Be by more than six-fold. In the troposphere however, production of ^7Be is significantly lower, and the residence time of ^7Be is much shorter (approx. 22-48 days) due to rapid washout [65, 66]. This results in a fairly high concentration gradient between the stratosphere and the troposphere, with tropospheric air generally containing 1-2 orders of magnitude less ^7Be (Bq m^{-3}) than in the stratosphere. ^7Be in stratospheric air sampled at 18-20 km ranged between 0.16

- 0.58 Bq m^{-3} while ^{7}Be in tropospheric air ranged from 0.005 to 0.02 Bq m^{-3} ^{7}Be [53, 67].

1.4.6 Stratosphere-troposphere exchange and seasonal variability

The activity of ^{7}Be in the stratosphere remains fairly constant. ^{7}Be concentrations in the troposphere and near- surface air and thus the amount of ^{7}Be available to ecosystems exhibit seasonal fluctuations [68, 69]. ^{7}Be concentrations in near surface air generally range between 0.001 - 0.007 Bq m^{-3} [69, 70]. Seasonal variations in ^{7}Be concentrations can be explained by four processes [69]: 1) Stratosphere-troposphere exchange; 2) vertical mixing within the troposphere; 3) air mass transport from middle to high latitudes; and 4) wet deposition.

Stratosphere-troposphere exchange increases ^{7}Be concentrations in the troposphere and near-surface air. The high concentrations of ^{7}Be measured in the USA are correlated with stratospheric air masses as indicated by potential vorticities [71]. Maximum mixing between stratosphere and troposphere occurs in spring at mid-latitudes, and generally higher concentrations of ^{7}Be are measured in the surface air at this time [54, 69]. Intense thunderstorms may also mix stratospheric air downward, thus increasing the amount of ^{7}Be available for scavenging by precipitation [72]. In summer due to the warming of the earth's surface, convection increases, which transports ^{7}Be from the upper troposphere to the near-surface air. Regions with a pronounced seasonal variation in rainfall amount show an inverse relationship between rainfall and ^{7}Be in the air, demonstrating that washout of ^{7}Be can have a significant impact on its surface air concentration [69].

1.4.7 Atmospheric deposition

Beryllium-7 can also reach the surface under clear sky conditions via dry deposition (particle sedimentation), but is only removed efficiently from the troposphere by precipitation scavenging (wet deposition). Experimental results show that around 90% or more of the total ^{7}Be deposition in temperate zones generally takes place via wet deposition [42, 73, 74, 75, 76]. Wet deposition of ^{7}Be occurs through both below-cloud scavenging (washout) and in-cloud scavenging (rainout). Washout occurs during early stages of precipitation and quickly depletes the lower troposphere of ^{7}Be-bearing aerosols. Rainout delivers ^{7}Be from within the cloud layer to the earth's surface and is active throughout the duration of a precipitation event. Serial sampling within storms shows that specific activities of ^{7}Be in precipitation decrease sharply in the beginning of the storm event and then remain fairly constant suggesting a change from washout to rainout as the predominant wet deposition process [42, 75, 77]. The annual deposition of ^{7}Be worldwide ranges from 400 to 6500 Bq m^{-2} and is related to rainfall (Table 3). Low annual depositions of ^{7}Be have been reported for sites in the Middle East and Mediterranean regions, as well as at East Antarctica. The highest annual deposition of ^{7}Be is reported for high rainfall areas in New Zealand.

Table 3: Annual atmospheric deposition of ^{7}Be and rainfall at different locations

Location	Latitude	^{7}Be Deposition (Bq m^{-2})**	Rainfall (mm)	Period	Reference
Malaga, Spain	36° N	412	308	1992-1999	[78]
Thessaloniki, Greece	40° N	736	424	1987-1992	[75]
Roskilde, Denmark	55° N	738	564	1990-1993	[79]
Bavaria, Germany	49° N	990	616	1989	[80]
Heidelberg, Germany	49° N	1250	810	1960-1961	[80]
Canberra, Australia	35° N	1030	660	1988-1989	[42]
Geneva, Switzerland	46° N	2095	966	1997-1998	[81]
Galveston TX, USA	29° N	2451	1167	1989-1991	[32]
Oak Ridge TN, USA	36° N	2017	1251	1982-1984	[34]
Bombay, India	19° N	1262	2277	1955-1970	[82]
Hokitika, New Zealand	42° N	6350	2634	1985-1986	[83]
East Antarctica	70° N	700	n.a.	n.a.	[84]
North Pacific and Atlantic oceans	0-60° N	1290	n.a.	n.a.	[85]

**Uncertainties were not denoted in the respective publications

Beryllium-7 depositional fluxes at various locations show a high positive correlation with rainfall, and temporal variations in the depositional flux of this radionuclide generally follow local rainfall patterns [34, 75, 76, 78, 80]. Average dry deposition velocities of ^{7}Be at different locations, including the Pacific and Atlantic oceans, are reported to be in the range of 0.004 to 0.074 m s^{-1} [75, 85]. Since some anthropogenic atmospheric radionuclides such as ^{90}Sr and ^{137}Cs are also found attached to submicron-sized aerosols, the deposition velocity of ^{7}Be may be a useful parameter in estimating the long-term deposition of radioactive pollutants from remote sources [83].

1.4.8 Distribution in the freshwaters

Partitioning coefficients ($K_d = [\text{Bq kg}^{-1}]/[\text{Bq L}^{-1}]$) for ^{7}Be between river water and various substrates were determined in the laboratory. It was found that most mud, silt, and clay minerals have a K_d near 10^5 L kg^{-1} under neutral to alkaline conditions (pH > 6) [86, 87]. Values of K_d in the freshwaters were reported to be greater than 10^4 [62, 88, 89]. This shows that ^{7}Be strongly sorbs to the fine particles and clay minerals in soils.

^{7}Be mobility may be enhanced in natural waters by the formation of soluble fluoride (F) and organic acid complexes [86]. Despite numerous studies documenting the production of ^{7}Be and the flux to the surface of the earth, there are relatively few studies documenting its mobility and export from watersheds. ^{7}Be inventories measured in the lake sediment cores were from 35- 875 Bq m^{-2} with ^{7}Be activities in the upper few centimetres of cores noted from 0.02 - 0.55 Bq g^{-1} [48, 90]. A strong seasonality for ^{7}Be fluxes in sediment traps in Lake Zurich was found, with maxima in July and August [91].

1.4.9 Distribution in the vegetation and soils

Beryllium-7 is delivered to ecosystems primarily as Be^{2+} in slightly acidic (pH<6) rainfall. The Be^{2+} ion is competitive for cation exchange sites because of its high charge density. As ^{7}Be^{2+} ions come in contact with soils and vegetation, it is rapidly accumulated by exchange surfaces. Forest canopies may decrease the amount of ^{7}Be that reaches soils and streams [92]. After deposition, most ^{7}Be decays in the soils, but some may be exported in particulate or dissolved form. Some areal activity densities of ⁷Be in soils and grasses, which have been reported in the literature, are summarised in Table 4. Temporal variations of ⁷Be areal activity densities at undisturbed sites are likely to occur where seasonal rainfall variation exists [30].

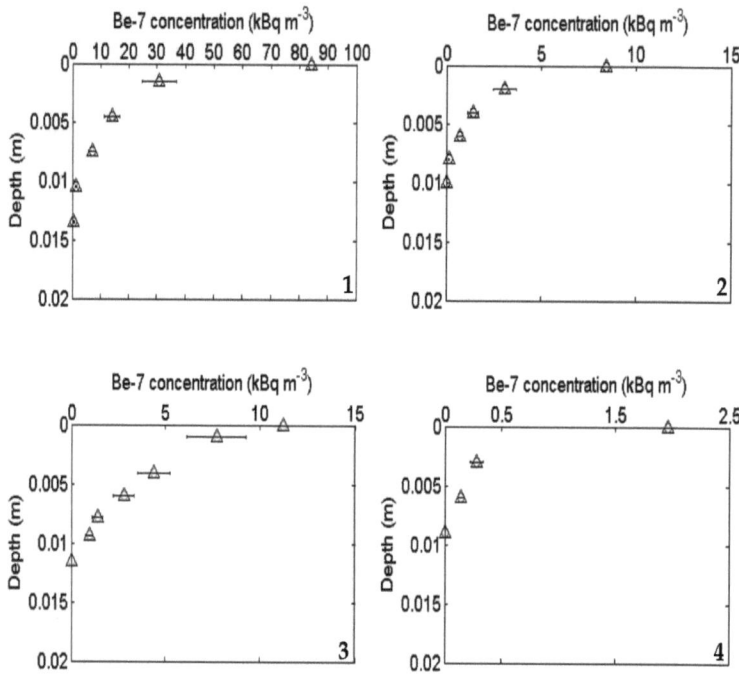

Figure 7: Depth distribution of ⁷Be at different study sites. Figures 1, 2, 3, and 4 are modified by the author from the sources in the literature [30], [33], [19], [94] respectively.

⁷Be does not penetrate deep into the soil profile as it is sorbed after it is deposited at the soil surface. The activity concentration of ⁷Be in soils generally shows an exponential decrease [30, 35, 93]. The depth at which activity concentration decreases to the half of its initial value is called the penetration half-depth. The half depths of

this radionuclide range from 0.4 to 3.7 mm [30]. In unsaturated marsh soils ^{7}Be can be found at a depth of 100 mm [34]. Particles mobilised by the infiltration of rainwater and transported through small cracks in the soil surface formed during relatively dry periods accounts for this finding. ^{7}Be penetration in the soils is primarily controlled by physical properties such as vegetation cover, soil density, and structure [30, 42]. An example of ^{7}Be depth distributions can be seen in Figure 7.

Table 4: Terrestrial inventories of ^{7}Be

Sampling date	Location	Latitude	Analysed sample	Total ^{7}Be inventory (Bq m^{-2})	References
07/1982	Delaware, USA	39° N	Marsh + Grass	207 ± 27	[34]
07/1984	Oak Ridge TN, USA	36° N	Grass + Soil	673 ± 22	[34]
01/1985	Wallops Is. VA, USA	38° N	Vegetated march soil	673 ± 48	[34]
01/1985	Wallops Is. VA, USA	38° N	Unvegetated march soil	107 ± 19	[34]
09/1988	Black Mt, Australia	35° N	Grass + Soil	202 ± 57	[30]
09/1988	Black Mt, Australia	35° N	Bare soil	135 ± 9	[30]
09/1989	Black Mt, Australia	35° N	Grass + Soil	400 ± 144	[30]
09/1989	Black Mt, Australia	35° N	Bare soil	156 ± 42	[30]
09/1989	Black Mt, Australia	35° N	Grass + Soil	205 ± 105	[30]
09/1989	Black Mt, Australia	35° N	Bare soil	95 ± 9	[30]
01/1991	Bologna, Italy	44° N	Grass + Soil	198 ± 9	[33]
01/1991	Bologna, Italy	44° N	Bare soil	125 ± 8	[33]
03/1991	Bologna, Italy	44° N	Grass + Soil	157 ± 8	[33]
03/1991	Bologna, Italy	44° N	Bare soil	153 ± 13	[33]
05/1996	Idaho, USA	44° N	Grass + Soil	139 ± 22	[62]
01/1998	Crediton, UK	50° N	Bare soil	512 ± 10	[33]
06/1998	Maine, USA	45° N	Forested soil	165 ± 66	[95]
10/1998	Silverton Mill, UK	50° N	Bare soil	283 ± 26	[33]
12/1998	Maine, USA	45° N	Bog core + vegetation	554 ± 144	[95]
05/1999	Treynor IA, USA	41° N	Bare soil	121 ± 21	[35]
1997-2000	Taiwan	25° N	Grass + Soil	3280 ± 1738	[55]

1.5 ^{7}Be: a promising short-term soil erosion/deposition tracer

1.5.1 Literature survey and state of the art

Beryllium-7 has been used since the late 1990s to estimate soil erosion and deposition processes associated with individual periods of heavy rain at scales ranging from plots of a few square meters to fields of a few hectares.

Field and laboratory experiments were performed since the early 1990s suggesting that the initial vertical depth distribution of ^{7}Be mass activity density, Bq kg^{-1}, within the soil is characterized by a strong exponential decrease with depth, with most of the radionuclide being found within the upper few millimetres of the surface soil [18, 19, 21, 22, 30]. ^{7}Be is concentrated in the uppermost soil horizons (Figure 8) suggest that it can be used as a tracer for topsoil movement.

The successful use of ^{7}Be to document both the magnitude and spatial pattern of short-term (rainfall event-based) soil erosion/deposition on agricultural land and the associated rates has been reported in the UK [21, 33]. The approach used in this study was based on comparison of the ^{7}Be areal activity density, Bq m^{-2}, measured at a sampling point with a reference areal activity density where neither erosion nor deposition has occurred. Depletion of ^{7}Be areal activity density, relative to reference value, provides evidence of erosion, whereas areas of deposition are associated with increased areal activity densities. The main components of this empirical erosion/deposition estimation technique are presented in the study done in Chile [18].

The application of ^{7}Be technique was suggested in 2006 together with the anthropogenic radionuclide ^{137}Cs for distinguishing between sheet and rill erosion [21].

Until 2012 several studies were published on the ^{7}Be technique in Australia, UK, USA and Chile [18, 19, 21, 33]. All these studies use the empirical technique to estimate the erosion/deposition rates [33].

1.5.2 Key considerations of the current erosion/deposition estimation technique and their limitations

The current ^{7}Be technique uses several key assumptions to estimate the erosion/deposition rates. If these assumptions are not fulfilled the technique overestimates or/and underestimates the erosion/deposition rates. These assumptions along with the possible limitations are discussed in this section to highlight the improvements needed for the use of ^{7}Be in estimating soil loss.

Assumption 1

The relationship between the activity density of ^{7}Be and mass depth documented for the reference site is exponential and representative of the main sampled area. Mass depth must be used instead of linear depth while establishing the depth distribution of ^{7}Be in soils.

Limitation

It's needed to ensure that the soil properties, surface conditions and surface hydrology of the reference sites are essentially similar to those of the sampling area. Use of mass depth limits the importance of any minor contrasts in bulk density between reference sites and sampling sites.

Assumption 2

^{7}Be associated with the erosional event is spatially uniform.

Limitation

Assumption 2 is commonly met at the scale of the individual field. The spatial distribution of rainfall input and ^{7}Be fallout can be considered to be spatially uniform. But for the larger fields and watersheds there is a need to take into account the spatial variability of the radionuclide.

Assumption 3

Any pre-existing ^{7}Be is uniformly distributed across the area under investigation [33].

Limitation

This assumption is frequently difficult to meet. The necessary uniform spatial distribution of pre-existing ^{7}Be can generally found in four situations,

1. After a long dry period, when any pre-existing spatially variability of ^{7}Be is removed by radioactive decay.
2. After an extended period of low intensity rainfall that has not resulted in an erosion and soil redistribution and therefore redistribution of existing ^{7}Be fallout input.
3. After a field has been ploughed and the existing ^{7}Be has been mixed within the plough layer and the activity is below the level of detection.
4. The erosion events are separated by a period of sufficient length (e.g. two half-lives or ~106 days).

Assumption 4

The ^{7}Be deposited during an erosion event will be rapidly fixed by the soil particles and can only be redistributed by the mobilisation and redistribution of soil particles. There is no significant grain size selectivity in the mobilization and deposition of soil particles.

Limitation

This assumption has been widely confirmed by experimental investigations of fixation of ^{7}Be fallout inputs by soil particles. ^{7}Be is mainly associated with the finer soil particles [62]. If the selective removal of fines occurs, the amount of erosion may be overestimated. Similarly, if selective deposition of coarser particles with lower ^{7}Be activity occurs, the deposition is underestimated.

Assumption 5

The use of ^{7}Be for documenting soil redistribution rates is commonly restricted to bare soils.

Limitation

As surface vegetation cover readily fixes the ^{7}Be fallout, the proportion of it fixed by the soil will vary according to the density and nature of vegetation. On an agricultural field soil is covered with crops that limit this assumption to bare fields.

1.5.3 Scope of the further development of the technique for soil erosion/deposition quantification

From the literature survey done during this study it was found out that ^{7}Be used for erosion/deposition calculation was based merely on the field studies and the observed depth distributions of the radionuclide in soils. The available empirical relationship as explained in the previous sections assumes constant bulk densities in the soil column and uses several key approximations to estimate the soil redistribution. A mathematical model, which represents the transport of ^{7}Be in the soil, is necessary and does not exist to date.

The current erosion/deposition estimation approach using ^{7}Be is based on the steady state approximation and therefore it is limited to a single rainfall event. A systematic approach is necessary to develop a technique for non-steady state conditions to estimate erosion/deposition rates produced by several successive periods of heavy rainfall.

1.6 Research Objectives

The objectives of this thesis are: First, to construct a model based on the physical process of diffusion, to investigate the movement of ^{7}Be in soils; second to use this model to modify the existing soil erosion quantification technique from a single erosion event to multiple erosion events which are separated by short time intervals Δt and third, to investigate the scope of the assumptions of the ^{7}Be technique mentioned in section 1.5.2. To perform this erosion research a study site in Müncheberg, Germany is selected. Specific objectives of this thesis are addressed in the following.

1. The existing mathematical relationship uses an empirical exponential function, which includes mass depth (kg m^{-2}) and specific activity (Bq kg^{-1}) to describe the depth distribution of ^{7}Be in soils. This relationship is based on several key assumptions as discussed in details in the previous section. A detailed analysis of the validity of these key assumptions is not available as of today in the literature. One way of approaching these open challenges is the use of

mathematical tools such as differential equations. A 1-D diffusion model was constructed during this thesis to explore the migration of ^{7}Be in the soil with the use of atmospheric input and radioactive decay. Several mathematical simulations were performed before investigating the scope of the model for soil erosion quantification based on existing assumptions.

2. The existing erosion quantification methods compare the inventories of ^{7}Be at reference sites with those at the study plots. One important requirement of this approach is that the areal activity density of the radionuclide is essentially constant across the study site, before the erosional event under investigation had occurred. The necessary conditions to achieve this are to separate the events by a period of sufficient length (~106 days or 2 half-lives of ^{7}Be) and by ensuring that the rainfall during the preceding period did not cause significant soil redistribution and thus ^{7}Be redistribution. These conditions may be difficult to fulfil at study areas, where the annual precipitation is high and heavy rainfall frequently occurs during the wet season. In this thesis a numerical technique was used that is capable of simulating soil attached ^{7}Be redistribution produced by several successive periods of heavy rainfall. Thus the existing technique for erosion quantification using ^{7}Be will be modified from single erosion event to multiple erosion events.

3. Erosion/deposition estimation is dependent on the areal activity density of ^{7}Be at the reference sites. It is well documented that the erosion rates calculated with the use of reference inventory technique show ambiguous results due to the spatial variability of the radionuclide deposited at the soil surface [6, 23, 24, 25, 96]. An independent approach was developed for this thesis, which does not include the comparison of areal activity density at a sampling point with that at the reference sites.

4. Usually the maximum activity of the radionuclide is present in the upper 1-1.5 mm decreasing exponentially with depth [33]. The potential of using ^{7}Be to differentiate between rill erosion, interrill erosion and surface erosion was

considered to be a future aspect [18, 33]. This aspect of the ^{7}Be depth distribution is also explored in this thesis.

5. The existing ^{7}Be technique is suited only for bare soils [18]. However, the erosion rates will be estimated for the situations where the field was ploughed and was covered with vegetation.

2 VERTICAL TRANSPORT MODEL FOR ⁷BE IN SOILS

2.1 Introduction

Slow migration of radionuclides result in the availability of it for plant uptake. Similarly, if the migration process is fast the radionuclide can enter the groundwater table quickly. To understand the processes involved in the transport of radionuclides into the soil column it is necessary to predict their time-dependent behaviour.

Measurements of long-lived ^{137}Cs have been used in investigations of long-term (up to 40 years) water erosion in a wide range of environments [5, 28]. The redistribution of ^{137}Cs within the soil column represents a complex set of mechanisms including physical, physico-chemical and biological processes [25, 97]. To describe the migration and diffusion process of ^{137}Cs in undisturbed soil, transport models like the convection-dispersion model [38, 98], the compartment model [99] and other statistical distribution function models [100] have been employed.

In the case of ⁷Be, an empirical relationship based on a shape preserving exponential depth distribution function exists to date in the literature to describe the migration of ⁷Be in soils [18, 19, 21, 33]. It assumes constant bulk densities in the soil column and calculates the penetration depth of ⁷Be. This empirical approach does not consider the physical background while establishing the depth distribution of ⁷Be in soils and eventually calculating erosion/deposition rates.

As the first objective of this study a one-dimensional diffusion model was developed to understand ⁷Be behaviour in soils. Secondly, the model was used to estimate the short-term erosion rates for single and multiple rainfall event scenarios. The effective diffusion coefficient D (m² s⁻¹) and the input flux I_0 (Bq m² s⁻¹) were determined by fitting the model to the measured ⁷Be concentrations in the soil column.

2.2 One-dimensional ^{7}Be diffusion model

2.2.1 General assumptions of the model

Since Beryllium-7 concentrations are maximum in the surface layer [18, 19, 21, 33, 42] and decrease with depth, diffusion can be considered as a predominant movement process over the profile. The model is applied to different vertical distributions of ^{7}Be in soils. The parameter values estimated from this model represent the traces of physical and chemical processes involved in the ^{7}Be transport in the soil. The most important assumptions made in this model are:

1. Beryllium-7 is continuously deposited on the soil via wet and dry deposition.
2. There is no horizontal transport component of the model. This corresponds to the dominance of the vertical component.
3. The diffusion coefficient, D is considered constant over the soil column.
4. Diffusion coefficient D is constant over time.
5. The parameter D combines molecular diffusion and dispersion of the solute caused by geometry of porous structures, into a single constant. It takes into account soil porosity and tortuosity [101].

2.2.2 Model conceptualization

Taking into account diffusion and radioactive decay, ^{7}Be transport in soil can be described by Fick's law (1) and the continuity or conservation equation (2).

$$J(z,t) = -D\left(\frac{\partial C(z,t)}{\partial z}\right) \qquad (1)$$

$$\frac{\partial C(z,t)}{\partial t} = -\frac{\partial J(z,t)}{\partial z} - \lambda C(z,t) \qquad (2)$$

To solve the differential equation (2) following initial and boundary conditions in semi-infinite space and time domains $z, t \in [0, \infty)$ were considered. The solution is finite which implies that $C(z \to \infty, t) \to 0$.

Initial Condition:

$$C(z > 0, 0) = 0 \qquad (3)$$

Boundary conditions:

$$-D\frac{\partial C(z,t)}{\partial z}\bigg|_{z=0} = I_0 \quad \text{(Continues flux boundary condition)} \quad (4a)$$

$$J\big|_{z=0,t} = I\,\delta(t) \quad \text{(Pulse boundary condition)} \quad (4b)$$

The boundary conditions (4a) and (4b) describe continues flux of ^{7}Be at the soil surface and a short pulse of ^{7}Be-carrying rainfall deposited at the soil surface respectively.

In equations (1), (2), (3), (4a) and (4b) following terminologies have been used: z: depth in soil (m); z=0: soil surface; t: migration time in seconds; C (z, t): total volumetric concentration, $I_0(z, t)$: atmospheric flux of ^{7}Be at the soil surface, Bq m^{-2} s^{-1}; D: diffusion coefficient, m^2 s^{-1}; λ: radioactive decay constant, s^{-1}, representing a sink $S_i = -\lambda C(z,t)$; I: deposition density, Bq m^{-2}; $\delta(t)$: Dirac delta function. Equation (4a) represents a flux-boundary condition.

The differential equation (2) is solved for the boundary conditions (4a) and (4b) respectively.

2.2.3 1-D diffusion equation with flux-boundary condition

Using (1) in (2) and using the initial and boundary conditions defined by (3) and (4a) differential equation (5) was solved. The main steps of calculations using Laplace transform are explained in Appendix B (Part 1).

$$\frac{\partial C(z,t)}{\partial t} = -D\frac{\partial}{\partial z}\left(\frac{\partial C(z,t)}{\partial z}\right) - \lambda C(z,t) \quad (5)$$

The concentration of ^{7}Be in the Laplace space is written as:

$$\overline{C} = \frac{I_0}{s\cdot\sqrt{s+\lambda}\cdot\sqrt{D}} e^{-\sqrt{\frac{s+\lambda}{D}}z} \quad (6)$$

By rearranging (6) and after applying an inverse Laplace transformation as shown in Appendix C, we get the solution of (5).

$$C(z,t) = \frac{I_0}{2\sqrt{D\lambda}} \left[e^{-\sqrt{\frac{\lambda}{D}}z} \, \text{erfc}\left(\frac{z}{2\sqrt{Dt}} - \sqrt{\lambda t}\right) - e^{\sqrt{\frac{\lambda}{D}}z} \, \text{erfc}\left(\frac{z}{2\sqrt{Dt}} + \sqrt{\lambda t}\right) \right] \qquad (7)$$

Equation (7) gives the solution of the 1-D diffusion equation with radioactive decay as the potential sink of ^{7}Be in the semi-infinite soil column. The analytical solution expressed by (7) is illustrated with the help of set of input data to understand the behavior of ^{7}Be concentration distribution in Figure 8.

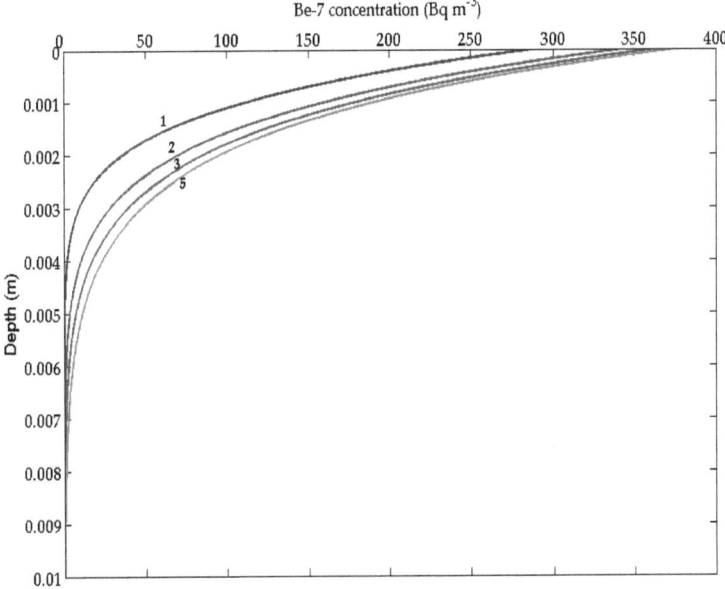

Figure 8: ^{7}Be concentration distribution for flux boundary condition in a semi infinite column of soil. The numbers on the curves indicate the time passed in half-lives of ^{7}Be.

Typical input data chosen for this purpose are $I_0 = 7.26$ Bq m^{-2} day^{-1}, $D = 2.85 \times 10^{-7}$ m^2 day^{-1}. The concentrations of ^{7}Be increase with increasing time. It also shows that the radionuclide penetrates exponentially in the soil with time.

Steady state depth distribution of ⁷Be in soil

For the steady state solution of ⁷Be in soils we used $t \to \infty$ in (7),

$$C(z, t \to \infty) = \frac{I_0}{\sqrt{D\lambda}} e^{-\sqrt{\frac{\lambda}{D}} z} \qquad (8)$$

From Equation (8) and the simulations shown in Figure 8 it is concluded that the ⁷Be transport in soil follows an exponential depth distribution if the governing processes are diffusion and radioactive decay. Empirical distributions used elsewhere in the literature [21, 33] are confirmed using a differential equation approach in this study. Both the unknown parameters D and I_0 are evaluated by fitting Equation (8) to the depth distributions established during this study for ⁷Be in soils and the data available in the literature [18, 19, 21, 33, 42].

2.2.4 1-D diffusion equation for Pulse boundary condition

For the situations where ⁷Be is instantaneously deposited on the soil surface by a heavy rainfall event, boundary condition (4b) is used to solve the differential equation (5). For such situations the concentrations and the flux of ⁷Be given by can be written as: $C_0(z) \equiv 0$; $J_0(t) = I(z=0, t) \, \delta(t)$

Using $C_0(z)$ and $J_0(t)$ and (4b) and applying a Laplace transform we get the concentration of ⁷Be in Laplace space (The main steps of the calculation are given in Appendix B (Part 2)):

$$\overline{C} = \frac{I}{\sqrt{D} \cdot \sqrt{s+\lambda}} e^{-\sqrt{\frac{s+\lambda}{D}} z} \qquad (9)$$

Using inverse Laplace transformation of (9) and rearranging we get the solution of equation (5) written as:

$$C(z, t) = \frac{I}{\sqrt{\pi D t}} e^{-\frac{z^2}{4Dt}} \cdot e^{-\lambda t} \qquad (10)$$

Equation (10) is a well-known solution for pulse-like input and has been used for ¹³⁷Cs migration studies in soils after the Chernobyl accident [101, 102, 103].

The movement of ^{7}Be in the depth after a short pulse of rainfall event is being shown in Figure 10 for different migration times. Parameter values chosen in this simulation are I_0 = 7.26 Bq m^{-2} day^{-1} and D = 2.85 × 10^{-7} m^2 day^{-1}. At t=10 days in Figure 9, ^{7}Be concentration has decreased due to diffusion and radioactive decay.

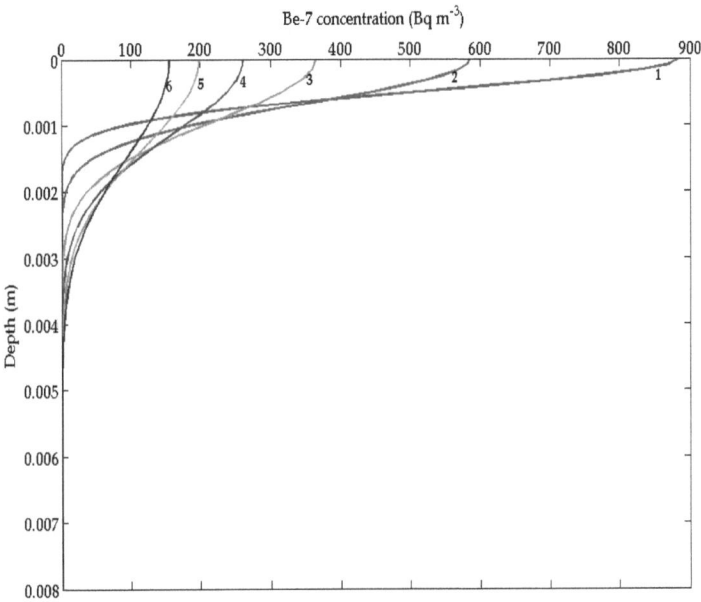

Figure 9: Depth distribution of ^{7}Be for different migration times. 1, 2, 3, 4, 5, 6 on the curves indicate the migration times 0, 5 days, 10 days, 20 days, 30 days and 40 days respectively.

2.3 Total inventory of ^{7}Be at the reference sites

The total ^{7}Be inventory at reference sites in Bq m^{-2} (where no erosion/deposition occurs) is calculated by integrating (8) from 0 to ∞.

$$\int_0^\infty \frac{I_0}{\sqrt{D\lambda}} e^{-\sqrt{\frac{\lambda}{D}}z} = \frac{I_0}{\lambda} = A_{Ref} \qquad (11)$$

2.3.1 Time evolution of total inventory and depth distribution of ⁷Be at a eroded position

The time it takes to approach the equilibrium depth distribution and total inventory of ⁷Be is crucial for estimating soil erosion rates. If the time period between two erosion events, Δt, is small the equilibrium technique using reference sites overestimates erosion rates. Thus it is necessary to calculate the time ⁷Be takes to achieve its equilibrium concentration after an erosion event. This is done in two parts. In the first part the total inventory evolution is calculated and in the second part the depth distribution evolution is considered for an eroded depth of magnitude Δz.

Part I: Time evolution of total inventory of ⁷Be after erosion

The change in concentration of ⁷Be, C (z, t) (Bq m⁻³) with time t (days) can be expressed as:

$$\frac{dC(t)}{dt} = I_0 - \lambda C(t) \tag{12}$$

If an erosion event occurs at time t = 0 and a small layer of soil Δz is lost, assuming that equilibrium had been achieved by ⁷Be, the total inventory of ⁷Be after erosion A(t₀), is calculated by integrating (8) from z= Δz to ∞.

$$A(t_0) = \int_{\Delta z}^{\infty} C(z, t \to \infty) dZ = \frac{I_0}{\lambda} \cdot e^{-\sqrt{\frac{\lambda}{D}} \cdot \Delta z} \tag{13}$$

ΔZ is the eroded layer of soil in meters.

The time evolution of the concentration of ⁷Be from time t=0 to t = t' is calculated by integrating (13) from 0 to t' (Appendix B (Part 3)):

$$A(t') = \frac{I_0}{\lambda}\left[1 - \left(e^{-\lambda t'} - e^{-\lambda t'} \cdot e^{-\sqrt{\frac{\lambda}{D}} \cdot \Delta z}\right)\right] \tag{14}$$

Here A (t') is the total inventory (Bq m⁻²) of ⁷Be evolved at t=t' after an erosion event at t=0. Imposing $t \to \infty$ on (14), $A(t') = \frac{I_0}{\lambda} = A_{Ref}$

This corresponds to (11) for the ^{7}Be inventory at reference sites. It suggests that as the time goes on, eventually all the radionuclide based information about erosion event is lost. Total inventory evolution with different half-lives of ^{7}Be is shown in Table 5. $A(t')$ and A_{ref} are estimated by using equations (11) and (14). It is clear from the Table 5 that the total inventory at an eroding position requires minimum three half-lives of ^{7}Be to achieve equilibrium concentration (90-95%). Systematic simulations for the total inventory evolution in time are discussed in the next sections of this chapter.

Table 5: Total inventory evolution of ^{7}Be at an eroding position

Eroded layer Δz (mm)	A(t')/Aref (%)	Half-lives of ^{7}Be
1	76	1
	88	2
	94	3
	97	4
	99	5
2	64	1
	82	2
	91	3
	95	4
	98	5

Part II: Time evolution of the depth distribution of ^{7}Be after erosion

With the use of depth distribution of the radionuclide in soil, it is possible to distinguish the contribution of soil coming from the surface and from the deeper layers. The equilibrium depth distribution of ^{7}Be is given by (8).

When a layer of soil, Δz is eroded at the study area the depth distribution after erosion, C_E, is written as follows and is shown in Figure 10.

$$C_E = \frac{I_0}{\sqrt{D\lambda}} e^{-\sqrt{\frac{\lambda}{D}}(z+\Delta z)} \tag{15}$$

Thus to investigate the evolution of the depth distribution of ^{7}Be, the initial and boundary conditions (A) and (B) are applied to (5)

Initial condition

$$C(z',t=0) = C_E \tag{A}$$

After erosion z is transformed to $z + \Delta z$ and is denoted by z' in (A)

Boundary Condition

$$-D\frac{\partial C(z,t)}{\partial z}\bigg|_{z=0} = I_0 \tag{B}$$

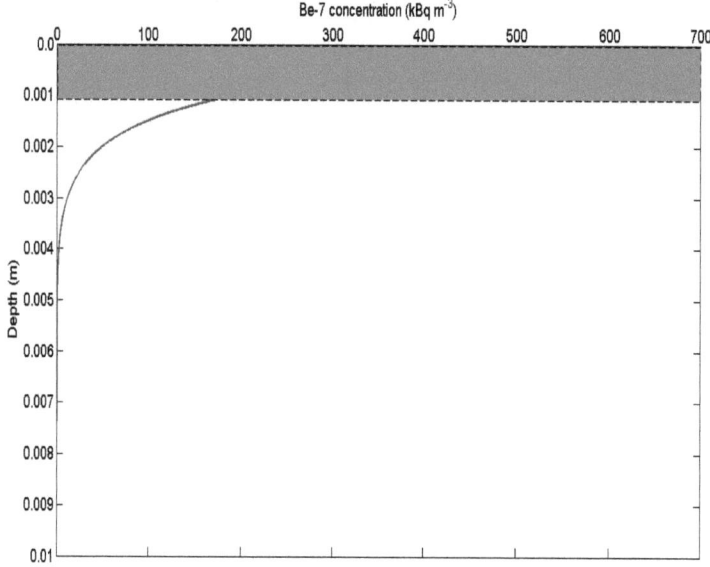

Figure 10: Hypothetical depth distribution curve for ^{7}Be concentration in soil. The shaded region indicates the eroded layer of soil Δz. The red curve shows the depth distribution of ^{7}Be after an erosion event.

Applying a Laplace transformation on (5) and using (A) and (B) we get the concentration of ⁷Be $\overline{C}$ in the Laplace space. After performing an inverse Laplace transform, equation (16) is obtained. C(z,t) in equation (16) gives the depth distribution of ⁷Be. The detailed mathematical approach to this problem is presented in Appendix B (Part 4).

$$C(z,t) = \frac{I_0}{2\sqrt{D\lambda}}\left[e^{-\sqrt{\frac{\lambda}{D}}z}\operatorname{erfc}\left(\frac{z}{2\sqrt{Dt}} - \sqrt{\lambda t}\right) - e^{\sqrt{\frac{\lambda}{D}}z}\operatorname{erfc}\left(\frac{z}{2\sqrt{Dt}} + \sqrt{\lambda t}\right)\right] + C_E \quad (16)$$

Equation (16) is a unique solution of (2) for the boundary condition given by (B). The solution (16) contains the concentration of ⁷Be left after an erosion event C_E and the function, which describes the vertical profile built up with the help of diffusion and radioactive decay.

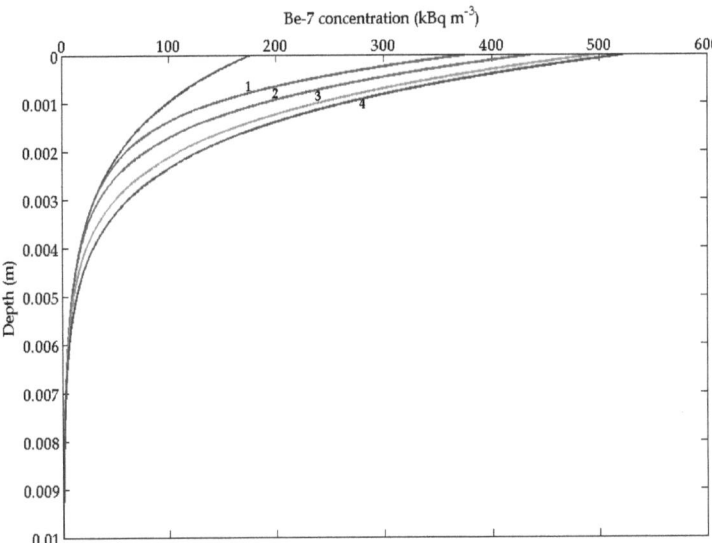

Figure 11: Time evolution of ⁷Be concentration with depth for four migration times. Blue curve: ⁷Be distribution left in the soil after and erosion event. 1, 2, 3 and 4 indicate four migration times in days t = 20, 40, 80, 150 respectively.

The plausibility tests for the equation (16) are given as:

1. For $t \to \infty$ equation (16) achieves the steady state solution given by (8),

$$C(z,t) = \frac{I_0}{\sqrt{D\lambda}} e^{-\sqrt{\frac{\lambda}{D}}z}(1+e^{-\sqrt{\frac{\lambda}{D}}z}); \text{ As } e^{-\sqrt{\frac{\lambda}{D}}z} < 1 \text{ and } \text{erfc}(-\infty) = 2$$

$$C(z,t) \approx \frac{I_0}{\sqrt{D\lambda}} e^{-\sqrt{\frac{\lambda}{D}}z} \tag{17}$$

2. Using $t = 0$ in (16), it is identical to initial condition (A)

$$C(z,0) = C_E; \text{ as } \text{erfc}(\infty) = 0 \tag{18}$$

In Figure 11 evolution of $C(z, t)$ with time for an eroded depth is shown for four migration times. For this purpose a typical value of eroded depth $\Delta z = 1$ mm is used. Figure 11 shows that the depth distribution requires 3 half-lives of ^{7}Be to achieve steady state given by equation (8). This result is identical to the total inventory evolution given in Table 5.

2.4 Estimation of short-term erosion rates using ^{7}Be diffusion model

To quantify eroded layer Δz at the study site, two situations were considered, which use the steady state and the non steady state approach respectively.

2.4.1 Steady state approach

This approach is used if the total inventory and the depth distribution of ^{7}Be at the study sites and the reference sites have uniform distributions (steady state) before the erosion occurs. After an erosion event the inventory of ^{7}Be (Bq m^{-2}) at study sites is measured and compared to that at the reference site and eroded depth is calculated by following method:

If Δz is the layer of soil eroded at the study plot then the inventory of ^{7}Be at the sampling point after erosion can be calculated by integrating (9) from Δz to ∞.

$$A_s = \int_{\Delta z}^{\infty} C(z, t \to \infty) dz = \frac{I_0}{\lambda} \cdot e^{-\sqrt{\frac{\lambda}{D}} \cdot \Delta z} \tag{19}$$

by rearranging (19) and using (11) we get,

$$\Delta z = \sqrt{\frac{D}{\lambda}} \cdot \ln\left(\frac{A_{Ref}}{A_s}\right) \tag{20}$$

If Δz in (20) is negative it implies that erosion has occurred and vice versa.

2.4.2 Non steady-state approach

This approach is used for calculating the eroded depth Δz for multiple erosion events. In situations where the erosion events are separated by a short time interval Δt, the steady state approach is not applicable because of large differences between the reference inventory and the inventory of ^{7}Be at the measurement point. This difference is due to the loss of ^{7}Be by erosion at the sampling point after the first event and not enough time for the ^{7}Be to reach a steady state (~106 days [33]). For such situations we cannot compare the inventories of ^{7}Be at a study plot with that at the reference site.

2.4.2.1 Crank-Nicolson scheme

For non steady-state conditions the numerical Crank-Nicolson scheme was used to estimate the ^{7}Be inventory within short time intervals between two erosion events. This is a standard scheme to solve diffusion equation, as it is unconditionally stable and highly accurate. It uses a finite differences approach for the discretization of differential equations. Its application to the diffusion model given by equation (5) is given by equation C1. A detailed discussion on the Crank-Nicolson scheme is given in Appendix D.

$$-\frac{D}{2\Delta z^2}C_{i-1}^m + \left(\frac{1}{\Delta t} + \frac{D}{\Delta z^2} + \frac{\lambda}{2}\right)C_i^m - \frac{D}{2\Delta z^2}C_{i+1}^m$$
$$= \frac{D}{2\Delta z^2}C_{i-1}^{m-1} + \left(\frac{1}{\Delta t} - \frac{D}{\Delta z^2} - \frac{\lambda}{2}\right)C_i^{m-1} + \frac{D}{2\Delta z^2}C_{i+1}^{m-1} \tag{C1}$$

The MATLAB implementation of equation (C1) is given in Appendix F. To test the ability of this numerical scheme to calculate inventories of ^{7}Be in soil the following simulations were performed.

1 Comparison of numerical and analytical solution

The Numerical solution was compared with the steady state solution and the truncation errors determined for the approximated solution. The transient behaviour of the solutions to equation (5) is smooth and bounded, meaning the solution does not develop local or global maxima that are outside the range of the initial data that is, the Crank-Nicolson scheme is unconditionally stable. The time increments (Δt) and the depth increments (Δz) were chosen to obtain the smallest difference between the numerical and analytical solution. The percentage error for this combination of M and N was < 1%. Depth and time increments, $\Delta z = 10^{-5}$ m and $\Delta t = 0.06$ day were estimated for t_{max} = 90 days and z = 2 cm. The simulations were performed for a number of time steps, M = 1000, 2000 and 2500 and a number of space steps, N = 500, 1000 and 2000 used for the solution matrix. For each combination of M and N, the numerical solution was compared with the analytical solution and absolute and percentage errors were determined. The best estimation for depth and time increments for the minimum error was N = 2000 and M = 1500.

2 Truncation error analysis for the Crank-Nicolson scheme

The truncation error for the Crank-Nicolson scheme is $\sigma(\Delta z^2) + \sigma(\Delta t^2)$, where σ represents the rate at which the truncation error goes to zero.

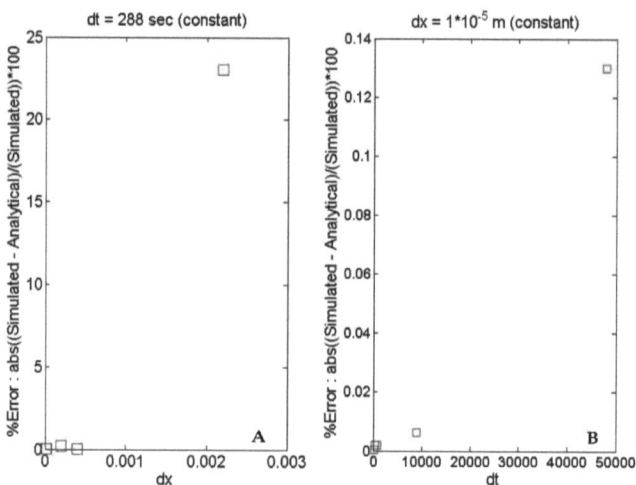

Figure 12: A: Truncation errors for Crank-Nicolson scheme as a function of Δx for fixed Δt. B: Truncation errors for Crank-Nicolson scheme as a function of Δt for fixed Δx.

The true magnitude of the Truncation Error (TE) is given as:

$$TE = K_t \Delta t^2 + K_x \Delta z^2$$

K_t and K_x are constants that depend on the accuracy of finite difference approximations.

To make TE arbitrarily small, both Δz and Δt must approach zero. Figure 12 gives the comparison of truncation error for Crank-Nicholson scheme for fixed time and space steps, Δt and Δx. For a fixed time step Δt, figure 12A shows that the dependence of a truncation error for Crank-Nicolson scheme is nearly identical except for small Δx. Figure 12B shows how the truncation error of the Crank-Nicolson codes varies when Δt is reduced and Δx is fixed. The solutions were obtained for diffusion coefficient, $D = 3.65 \times 10^{-13}$ m² s⁻¹, depth of the soil column L= 2 cm and t_{max} = 90 days. The Crank-Nicolson scheme can obtain solutions for much larger Δt as it is unconditionally stable. As $\Delta t \to 0$ for finite Δx, the TE of Crank Nicolson scheme approaches a constant value.

2.4.2.2 Erosion quantification using the Crank-Nicolson scheme

The erosion estimation technique for non steady state condition can be understood by considering two erosion events E1 (t_1) and E2 (t_2) occurring at time $t=t_1$ and $t=t_2$ (Figure 13).

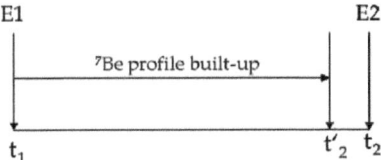

Figure 13: Scheme for the calculation of erosion rates for multiple erosion events. E1 and E2 are erosion events at time instances t_1 and t_2 respectively. t'_2 is the time instance before erosion event E2 occurs.

The necessary input parameters to solve the differential equation (5) using the numerical Crank-Nicolson method are as follows,

1. The initial condition is chosen as the inventory of ^{7}Be after the erosion event at t=t_1 C(z, t=0) =C_E as per equation (19). C_E is obtained from the measurement of ^{7}Be inventory at a sampling point after an erosion event.
2. The input flux I_0 (Bq m^{-2} s^{-1}) is estimated from the reference inventory by using equation (11).
3. The diffusion constant (m^2 s^{-1}) is evaluated by fitting equation (8) to the depth distribution of ^{7}Be at the reference sites.

Using the three input parameters mentioned above, depth distribution profile of ^{7}Be was numerically simulated until the time $t = t_2'$. The simulated profile at $t = t_2'$ of ^{7}Be is integrated over the depth to obtain the total inventory (equation (21)). The simulated total inventory at point $t = t_2'$ is denoted by A_{Num}.

$$A_{Num} = \int_0^{z_{End}} C_{Num}(z_i, t = t_2') dz_i \qquad (21)$$

here $z_i = (i-1)\Delta x$, $i = 1, 2, \ldots N$; N is the total number of spatial nodes including those at the boundary and Δx is the spacing between z_i. In equation (21), A_{Num} is the total inventory of ^{7}Be before erosion in Bq m^{-2} and C_{Num} is the simulated volumetric concentration of ^{7}Be in Bq m^{-3}.

$$\Delta z = \sqrt{\frac{D}{\lambda}} \cdot \ln\left(\frac{A_{Num}}{A_s}\right) \qquad (22)$$

The inventory of ^{7}Be obtained from (21) is compared to the inventory of ^{7}Be measured at the study site after erosion event E2 at $t = t_2$ and eroded/deposited layers are calculated by using (22).

3 NUMERICAL SIMULATIONS OF THE SYSTEM UNDER STUDY WITH THE ⁷BE DIFFUSION MODEL

As the first step to explore the scope of the mathematical model developed in this study, a range of numerical simulations were performed using equations (8), (10), (14) and (16) for known and estimated input values. Values of the parameters used with the simulations are presented in Table 6. These values are estimated by fitting the diffusion model (8) to the different depth distributions data documented at our study site in Müncheberg in the month of June 2011.

Table 6: Values of parameters taken for simulations

Parameter	Value
Diffusion coefficient (D) m²s⁻¹	$(3.65 \pm 0.75) \times 10^{-13}$
Input flux (I_0) Bq m⁻² s⁻¹	$(3.64 \pm 0.23) \times 10^{-5}$
Eroded depths (Δz) mm	0.5; 1; 1.5
Decay constant (λ) s⁻¹	1.51×10^{-7}
Depth of the soil column (L) m	0.01

The following simulations were performed before using the diffusion model to calculate eroded depths:

1. Time evolution of total inventory of ⁷Be for different eroded depths Δz.
2. Time evolution of depth distribution of ⁷Be for different eroded depths Δz.
3. Time evolution of depth distribution and total inventory of ⁷Be for varying input fluxes I_0
4. Evolution of depth distribution of ⁷Be for ploughing conditions.
5. ⁷Be distribution for a short rainfall pulse input.

3.1 Time evolution of the ⁷Be total inventory

To estimate the time required for the total ⁷Be inventory to achieve steady state, after a rainfall event with an eroded depth Δz, equation (14) was used. By using the parameter values given in Table 6, the time evolution of the ⁷Be inventory was calculated. Figure 14 shows the percentage of evolved inventory with time in days. Three curves represent the inventory approaching steady state (%) for different eroded depths. The evolution of the total inventory of ^{7}Be in time depends on the eroded depth Δz and the atmospheric input, J^0. The minimum time which total inventory (Bq m^{-2}) takes to reach the steady state (90-95%), is approximately 90-110 days. This suggests that for the steady state approximation, if the rainfall events are separated by a period less than 100 days erosion will be overestimated. This result confirms the limitation of the existing ^{7}Be technique for erosion estimation and is also noted elsewhere in the literature [18, 19, 21, 33].

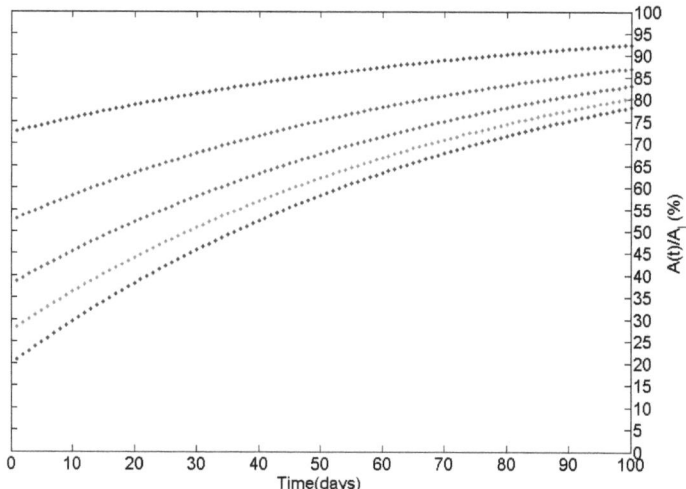

Figure 14: Time evolution of the total inventory of ^{7}Be for eroded depths Δz of 0.5 mm (Blue curve), 1 mm (Green Curve), 1.5 mm (Red curve) 2 mm (Cyan curve) and 2.5 mm (Violet curve). The curves start from an percentage inventory which is left after an erosion event.

A hypothetical situation for multiple erosion events is considered by assuming five erosion events of magnitude Δz mentioned in Table 6. These five events are simulated for steady state conditions in a period of approximately 400 days (Figure 15). This shows that if we want to use the steady state approach for multiple erosion event situations, we have to make sure that the multiple events are separated by a period of ~110 days (depending on the erosion rate) and within this time period there has not been any soil redistribution at the study plot. This is an impossible criterion to meet as natural conditions are dynamic and single rainfall events cannot be predicted.

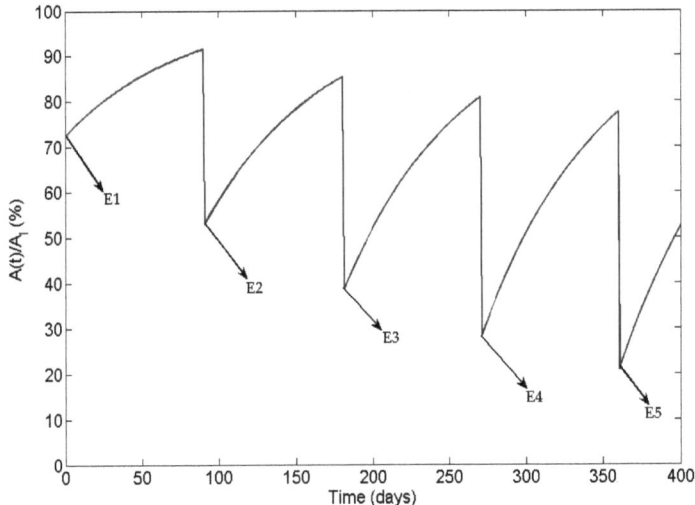

Figure 15: Time evolution of the total inventory of ^{7}Be for multiple erosion events E1, E2, E3, E4 and E5 with eroding depths of 0.5 mm, 1 mm, 1.5 mm, 2mm and 2.5 mm respectively.

3.2 Time evolution of the depth distribution of ^{7}Be

Along with the total inventory, the behaviour of the depth distribution was also studied. The results are presented in Figure 16. The concentration of ^{7}Be was simulated over the depth of 1 cm using Equation (16). Our simulations show that the depth distribution at the study site after an erosion event of takes a minimum of 100 days to reach steady state. For non-steady state conditions, e.g. when the time period between two erosion events is short (< 90 days) the deposited ^{7}Be concentration from the atmosphere is found at the soil surface (z=0).

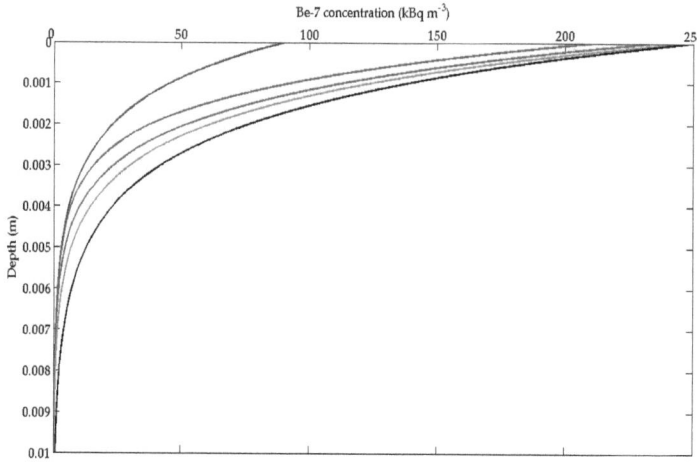

Figure 16: Time evolution of the ^{7}Be depth distribution after an erosion event with an eroded depth Δz = 1.5 mm. Blue curve: Depth distribution after an erosion event at t = 0; Green curve: depth distribution evolved at t = 35 days; Red curve: depth distribution evolved at t = 60 days; Cyan curve: depth distribution evolved at t = 80 days; Black curve: Steady state depth distribution which overlaps with the numerical depth distribution at t = 90 days.

3.3 Time evolution of the ^{7}Be total inventory and depth distribution for varying input fluxes I_0

While calculating erosion rates for multiple rainfall events, an assumption of constant input flux of ^{7}Be at the soil surface is considered. During the study period it was observed that the ^{7}Be input fluxes show seasonal variability. The total inventory and depth distribution evolutions for varying input fluxes are shown in Figure 17 and 18. Figure 17 shows that during the period of increasing input fluxes the total inventory reaches faster to the equilibrium state. If an input flux is doubled the total inventory reaches upto 110%. This leads to an overestimation of the erosion rate. The system is studied in more detail by using depth distribution for varying input flux condition and eroded depth of 1.5 mm. During this study it has been observed that atmospheric input is higher in the summer months (July-September) compared to the other period of the year. But, as seen at our study site, erosion events occurring during July-September are separated by short time intervals Δt. Therefore constant input flux is a good approximation.

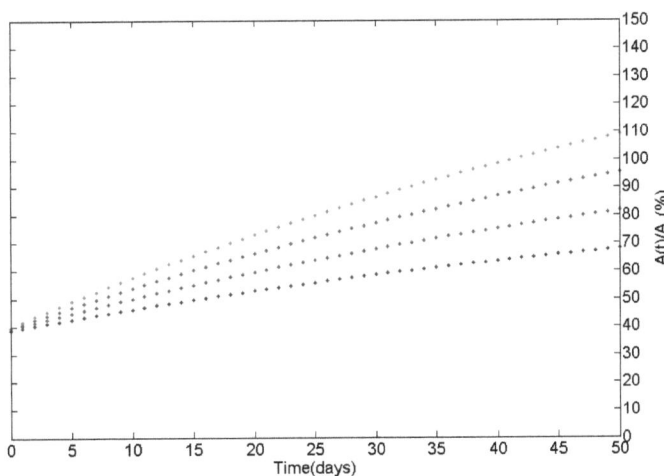

Figure 17: Time evolution of the total inventory of ^{7}Be for eroded depths Δz =1.5 mm for different input fluxes using Equation (14): 3.54×10^{-5} (Blue curve); 4.54×10^{-5} (Green curve); 5.54×10^{-5} (Red curve) and 7.54×10^{-5} (Cyan curve). The curves start from a 40% inventory which is left after an erosion event.

The depth distribution of ^{7}Be plotted in Figure 18 corresponds to input fluxes of 3.54×10^{-5} Bq m^{-2} s^{-1} and 6.54×10^{-5} Bq m^{-2} s^{-1} respectively. It can be seen from Figure 18 for higher input fluxes most of the ⁷Be concentration is found in the uppermost few millimetres of soil.

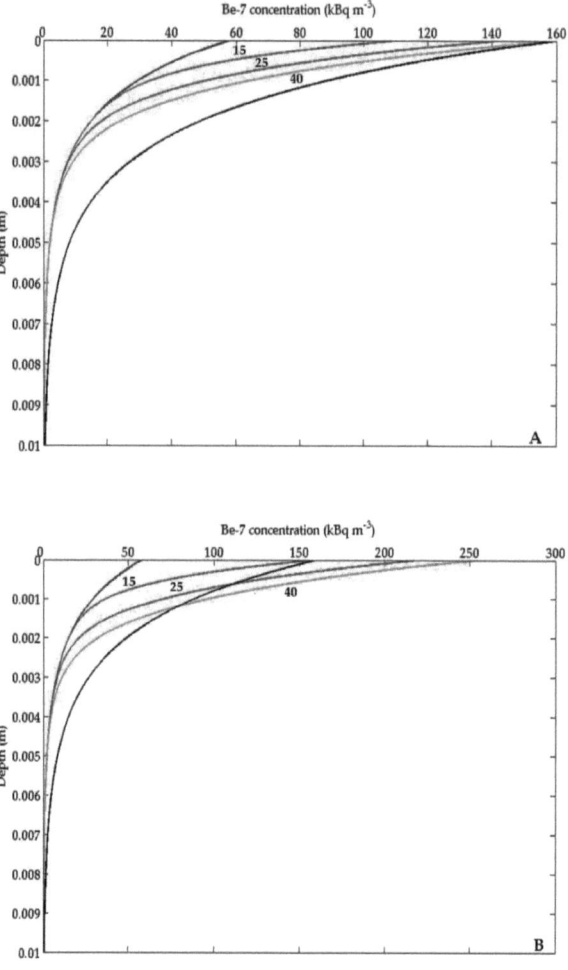

Figure 18: Time evolution of depth distribution of ⁷Be for varying input fluxes (A: I_0 = 3.54×10⁻⁵ Bq m⁻² s⁻¹, B: I_0 = 6.54×10⁻⁵ Bq m⁻² s⁻¹). The numbers on the curves indicate the simulation time in days. Blue curve: ⁷Be depth distribution after erosion; Black curve: steady state depth distribution of ⁷Be.

3.4 Time evolution of the ⁷Be depth distribution after ploughing

The study field at Müncheberg was ploughed two times a year. Due to the ploughing ⁷Be in the soil was mixed in the ploughing depth of 20-25 cm, meaning that the depth distributions were erased.

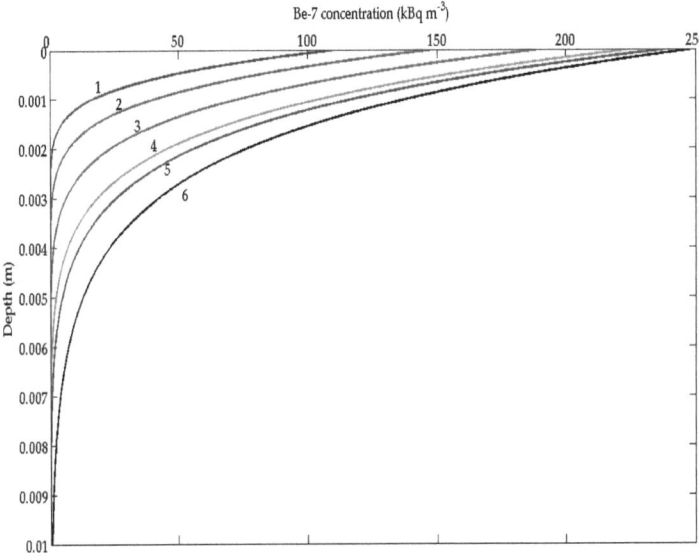

Figure 19: Evolution of the depth distribution of ⁷Be with time after ploughing. The numbers on the curves indicate the simulation times in days.

(1: 10 days, 2: 30 days, 3: 60 days, 4: 100 days, 5: 150 days and 6: 200 days)

To estimate erosion after ploughing we have numerically simulated several ⁷Be profiles at the study plot. For this purpose the ⁷Be depth distribution was simulated using the Crank-Nicolson scheme, for the initial condition given as $C(z, t=0)=0$ (Figure 19). The simulation results concluded that it requires 180-200 days for the ⁷Be depth distribution to reach steady state after ploughing. Therefore, to use the ⁷Be technique for erosion quantification after ploughing with the steady state approximation, a build-up period of 200 days must be taken into consideration.

3.5 ⁷Be distribution in soil for a pulse-like input

Short rainfall events deliver ⁷Be into soils. Such events are considered as pulse-like inputs. The differential equation (5) was solved using the Crank-Nicolson scheme with an instantaneous pulse input at the boundary at time t = 0.

The movement of ⁷Be into the soil column is being shown in Figure 20 for different simulation-times t. At t=10 days in Figure 20 the concentration of ⁷Be has decreased due to the physical process of diffusion and radioactive decay and is confined to the surface by sorption.

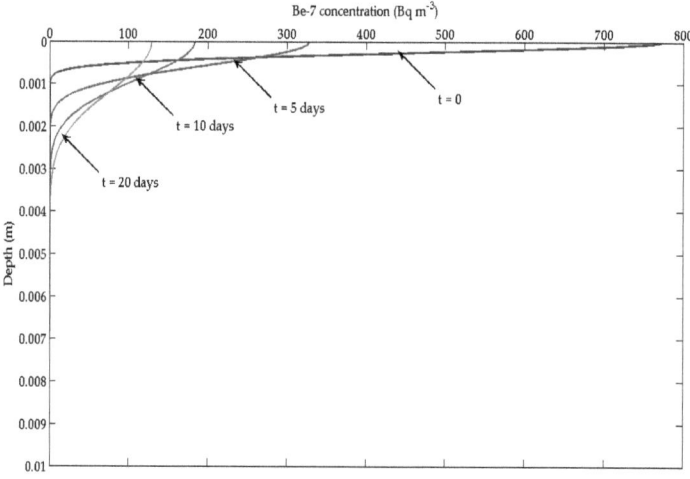

Figure 20: Depth distribution of ⁷Be for three migration times estimated with equation (10). The Blue curve is the pulse-like input at the boundary at t=0. The Green curve, Red curve and Cyan curve indicates the depth distribution at t = 5 days, t=10 days and 20 days respectively

3.6 Summary

A numerical simulation scheme was developed, using the Crank-Nicolson approach to solve the diffusion equation. The scope of the diffusion model was explored under pulse-input conditions to study the behaviour of ⁷Be for an instantaneous rainfall event. Before applying the model for erosion rates estimation several mathematical simulations were performed to understand the behaviour of ⁷Be for steady state and non steady state conditions. The simulations presented in this chapter are very illustrative and the non-steady state diffusion model developed here is essential for short-term erosion rates quantification.

4 FIELDWORK AND LABORATORY ANALYSIS

4.1 Introduction

The study-field is situated northeast from Berlin, Germany at Müncheberg. It belongs to the agricultural research institute, ZALF (Leibniz-Zentrum für Agrarlandschaftsforschung). It is specially designed to perform systematic study on slope-induced erosion caused by agricultural practices. Erosion quantification using the cosmogenic radionuclide ^7Be was conducted at this plot together with the erosion studies performed by ZALF.

4.2 Study site Müncheberg

Müncheberg is situated in the Märkisch-Oderland district in Brandenburg, Germany (52.6°N 14.3°E). The main research programmes carried at Müncheberg are monitoring programs in the agricultural landscape and long-term field studies.

The study plot was used under a "two plot utilization system" meaning that, the plot was divided into sub-plots, Plot1 and Plot 2, according to the mode of agriculture. Plot-1 is kept under "no-till" practices where the soil is scratched at the surface with minimal force of the machine that is specially designed for this purpose (Figure 21B). Plot-2 is ploughed with the conventional ploughing technique with ploughing depths reaching up to 25 cm (Figure 21A). In the winter months (October-February), Winter-Rye (Vital 10) was grown on both plots and in the summer months (June-September) Sudan grass (C4 plant) was grown. Both crops were harvested in the June and October. These harvested crops were used as fuel for bio-energy production [104].

Each sub-plot is 53.5 m × 6 m in area, with a slope angle of 6.2° (Figure 22 A). A tin board separates the two sub-plots made under different ploughing techniques. Sediment mobilised at the study plot by an erosion event was collected in the tin barriers and V-channels, which were installed at the bottom of the slope of each plot (Figure 22 and 23). Daily measurements for rainfall, wind speed and solar radiation, were obtained at the bottom of the plot by an automatic weather station.

Figure 21: Agricultural practices using different types of machines; A: No-till machines, B: Conventional ploughing machines

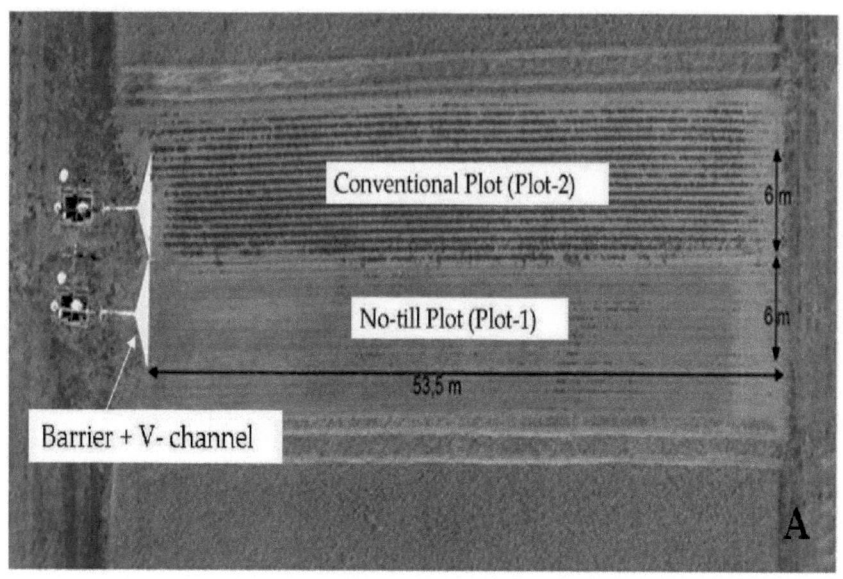
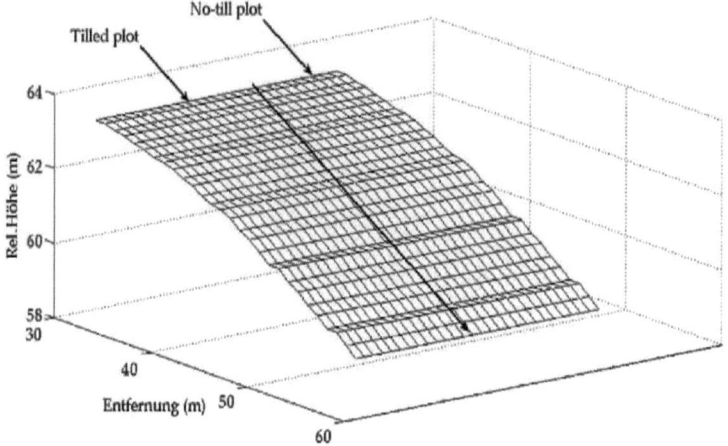

Figure 22: A: The experimental plot at Müncheberg research station. B: Schematic of the Müncheberg study plot. The bold arrow in the middle represents the direction of the slope.

Figure 23: The experimental plots with the barrier and funnel system at the bottom; A: barrier at no-till plot, B: barrier at till plot, C: Funnel system with automatic weather station.

4.3 Climate and soil characteristics at Müncheberg

The climate at the study site is a typical East-German continental inland climate. Winters at the study plot are cold with minimum temperatures of -15°C. In summer maximum temperatures up to 30°C were measured. From 2009-2011 the study site has experienced frequent long dry weather periods. The annual mean rainfall measured at Müncheberg from 1971-2010 is 525 mm and the annual mean temperature is 8.6 °C [105].

Table 7: Physical soil properties at the study site

Plot	Position	Geographical position		Bulk density (kg m^{-3})	Organic Matter (OM) (%)	Particle size (%)		
		Latitude (N)	Longitude (E)			*Sand	*Silt	*Clay
1	Top	52° 30' 56.5''	14° 7' 39.3''	1573	0.92	85	12	3
1	Middle	52° 30' 56.9''	14° 7' 39.8''	1674	0.98	84	13	3
1	Bottom	52° 30' 55.1''	14° 7' 40.7''	1578	0.97	82	14	4
2	Top	52° 30' 56.5''	14° 7' 39.3''	1525	1.03	85	12	3
2	Middle	52° 30' 56.9''	14° 7' 39.8''	1383	1.04	85	12	3
2	Bottom	52° 30' 55.1''	14° 7' 40.7''	1479	1.19	82	13	5

* Sand (> 63 µm), Silt (2 - 63 µm), Clay (< 2 µm)

The soil at the study site is a typical north-eastern German groundwater gray soil, which developed as a result of periodic water lodging by fluctuating groundwater tables. The soils at the study plot represent the subgroup brown calcareous subsoil and Luvisol. According to the German soil rating system, the study site mainly consists of sandy loam and sandy soils. The physical properties of the soil at the study plot are listed in Table 7. Soil structure was determined at three different parts of the plots. Dry densities were measured during this study on a monthly basis over a period of 24 months.

4.4 Sampling design

A systematic and non-stratified sampling design using a transect plot was chosen for this study. The coordinates were fixed using a high resolution GPS camera system. The transect was a continuous slope with measurement points 1 to 5, with 5m as the highest and 53 m as the lowest point (Figure 24). Uncertainties of the coordinates and the distances were in the range 2-5%.

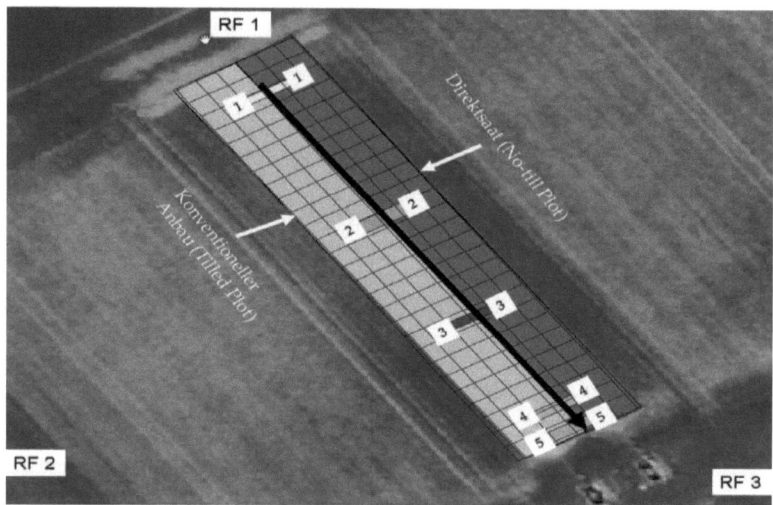

Figure 24: The sampling design for ^{7}Be measurements at Müncheberg with RF1, RF2 and RF3 representing the reference sites. The black arrow in the middle of the plot indicates the slope direction.

To determine event-based soil redistribution at the study plots a set of soils samples with five profiles (20 cm × 15 cm × 2 cm) located at 5m, 20m, 35m, 50m, 53 m were collected from each plot. Three reference sites were located at three sides of the plot (Figure 24). These reference sites represented a flat stable area with no visible evidence for the occurrence of erosion or sedimentation during rainfall events.

4.5 ^{7}Be measurement program for erosion/deposition quantification

The purpose of the study was to quantify soil erosion and deposition induced by multiple heavy rainfall events. Rainfall data was collected at the study plot using an automatic rainfall measurer. Most of the heavy rainfalls occurred during the period of July to early October. After the warm months from March to June the soil is dry and increasingly susceptible for water erosion. Thirteen samples were collected (5 at the tilled plot, 5 at the no-till plot and 3 at reference sites) over a period of 19 months (March 2010-October 2011) with a monthly or bi-monthly frequency, depending on the occurrence of rainfall events. Sampling was conducted for 3 different weather scenarios:

1. A rainfall event, which deposits soil on the tin barriers, placed at the bottom of the plot.
2. A rainfall event that redistributes the soil within the study plot and no soil is deposited at the bottom of the plot.
3. After dry periods with few rainfall events of small intensity or no rainfall events at all.

The rate of accumulation of ^{7}Be in the soil is likely to exceed the rate of loss by decay during the months of higher rainfall [94], while the ^{7}Be half-life (radioactive decay) will control the areal activity density during the months of lower rainfall. Sampling was performed during all seasons to study the seasonal variability of the total inventory of ^{7}Be in the soil.

Along with the 13 samples from the study plot, 9 samples were collected using a measuring cylinder to establish a depth distribution of ^{7}Be at the reference sites during different periods of the year. The sampling dates along with the rainfall data is given in the Figure 25. Sampling information is given in details in Table 8. The amount of rainfall data measured by the automatic weather station located at the study plot.

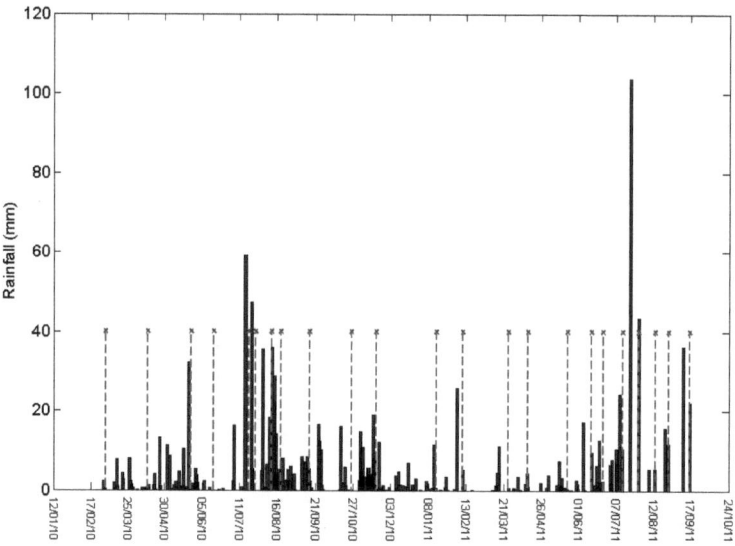

Figure 25: Rainfall events with soil sampling. Red dotted stems indicate the sampling events in 2010-11.

The particle size investigation of sediments collected in the tin barriers and V-channels was performed with the help of laser particle size measurements at the Asse-GmbH, Remlingen.

4.6 Sampling methodology

The sampling procedures and the availability of suitable sampling equipment are important potential constraints for the use of ^{7}Be in assessing soil redistribution. The choice of method for the collection of soil samples to determine levels of ^{7}Be is dependent on the purpose of the project, as well as the study site and its soil characteristics. For the analysis of the total number of soil samples the distribution of radionuclide in the soil and the sample mass is required. For this, soil samples from shallow depths need to be sampled because ^{7}Be occurrence is restricted to the immediate surface layer of the soil and shows a fast decrease with depth [18, 19, 21, 22, 33].

If the sample is not collected to a sufficient depth, it will not contain the full ^{7}Be inventory. If however the sampling depth is too great, inclusion of soil with distinguishing ^{7}Be activities may reduce the concentration of ^{7}Be in the overall sample to a low level or even to below the level of detection. Considering these criteria for measurement of ^{7}Be in the soil a sampling depth of 2 cm was chosen.

For characterizing the depth distribution of ^{7}Be at reference sites there is a need to cut the sample at a resolution of 1 mm and 2 mm depth increments. A cylindrical corer was developed in this study for measurement of the depth distribution.

4.6.1 Total inventory sampling using the Gauge-Scraper plate method

A scraper plate was developed during this work specifically for establishing the total inventory of ^{7}Be at reference and actual study sites. Sampling is done by two components: a gauge of dimensions 15 cm × 20 cm × 2 cm that can be placed on the ground to cut a specific part of soil and a scraper plate to cut a layer of soil of 2 cm.

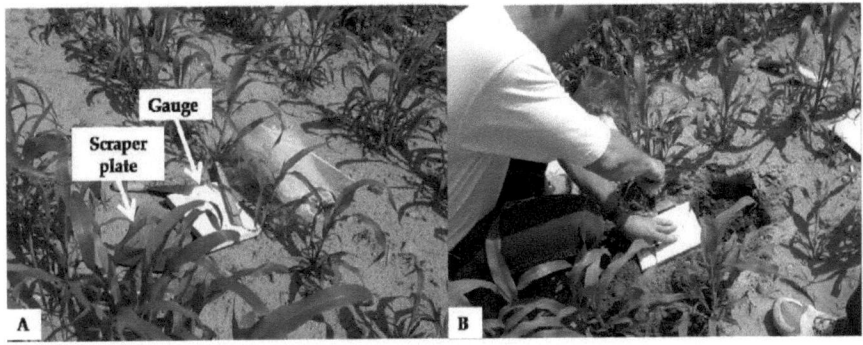

Figure 26: A: Gauge and scraper plate at the study site. B: The gauge is placed on the soil and with the help of a scraper plate the soil of a particular depth is cut.

Table 8: Sampling chart for soil samples at the site ZALF, Müncheberg

Sampling month	Date of rainfall event	Sampling date	Rainfall (mm)	Total rainfall duration (hrs)	Kinetic Energy of the raindrops (kJ m^{-2})*
March 2010	02.03.10	03.03.2010	3.7	25	32.4
April 2010	10.04.10	14.04.2010	13.2	1	179.9
May 2010	24.05.10	26.05.2010	32.3	10	764
June 2010^P	07.06.10	16.06.2010	2.4	7	31.0
July-1 2010	17.07.10	20.07.2010	59.1	9	1332.5
July-2 2010	23.07.10	27.07.2010	52.7	34	1031.4
August-1 2010	09.08.10	11.08.2010	62.4	6	700
August-2 2010	18.08.10	20.08.2010	51	20	1134
September 2010	14.09.10	16.09.2010	28.4	15	391.6
October 2010^P	20.10.10	26.10.2010	6.3	18	91.6
November 2010	10.11.10	19.11.2010	15.1	22	75
January 2011	13.01.11	17.01.2011	11	21	144
February 2011	04.02.11	09.02.2011	26	49	357
March 2011	17.03.11	22.03.2011	11	28	133
April 2011	13.04.11	15.04.2011	4.3	22	44.3
May 2011	14.05.11	23.05.2011	7.3	5	114.2
June 2011^P	22.06.11	27.06.2011	22.5	6	292.6
July 2011	14.07.11	18.07.2011	104	6	1762
August-1 2011	25.07.11	02.08.2011	22.7	52	524
August-2 2011	21.08.11	23.08.2011	5.4	2	113
September-1 2011	27.08.11	02.09.2011	11.7	4	209
September-2 2011	11.09.11	16.09.2011	36	5	944
October 2011^P	05.10.11	12.10.2011	5.1	4	21

^{P} Months where the study plot has been ploughed,* The energy at which the raindrops strike the rainfall sensor

At the sampling point a hole was made larger than the sampling depth. The gauge is put on the ground near the corner of the hole. The upper layer of soil of 2cm in height is taken as a sample with the help of the scraper plate (Figure 26).

Figure 27: A: Reference site RF3 at Müncheberg. B: Grass covered reference site, RF3

A household knife along with the scraper plate was used to cut the grass roots so that the scraper plate could sample the soil (Figure 27). Detailed information on the study sites and reference sites with their geographical coordinates is listed in Table 9.

Table 9: Description of the study site and reference sites

Measurement point	Location	Geographical position		Soil type
		Latitude (N)	Longitude (E)	
5 m	Till & No-till plot	52° 30′ 56.4″	14° 7′ 39.4″	Cultivated soils
20 m	Till & No-till plot	52° 30′ 56.9″	14° 7′ 39.8″	Cultivated soils
35 m	Till & No-till plot	52° 30′ 55.3″	14° 7′ 40.3″	Cultivated soils
50 m	Till & No-till plot	52° 30′ 55.2″	14° 7′ 40.7″	Cultivated soils
53 m	Till & No-till plot	52° 30′ 55.1″	14° 7′ 40.7″	Cultivated soils
RF1	Reference site 1	52° 30′ 57.0″	14° 7′ 39.8″	Bare soil + grass
RF2	Reference site 2	52° 30′ 55.4″	14° 7′ 38.2″	Soil+ long grass
RF3	Reference site 3	52° 30′ 54.2″	14° 7′ 40.2″	Soil+ grass+ small plants

4.6.2 Fine increment soil sampler for depth distribution measurements

To determine the depth distribution of ^{7}Be at the reference sites 9 cores were collected and cut into slices of 1 mm or 2 mm for establishing the depth distribution of ^{7}Be in soils.

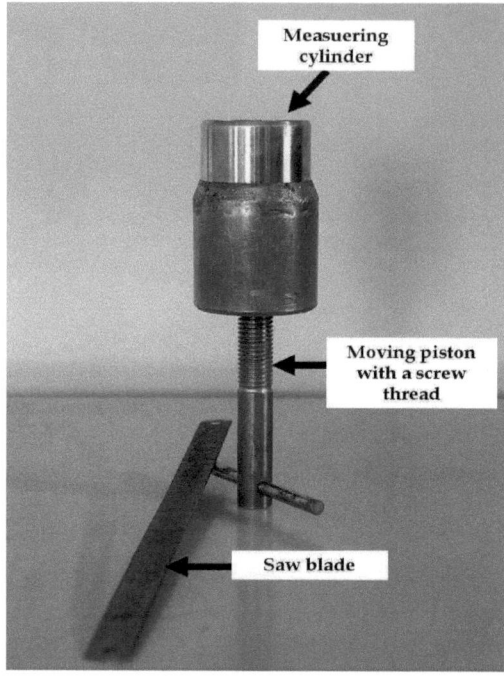

Figure 28: Fine increment cylindrical soil sampler (Photo: Klaus Schmidt)

A special device was developed for this study to facilitate slicing the cores into appropriate slices, based on the one described elsewhere [18]. This device comprised of, a measuring cylinder (5 cm diameter and 5 cm in height) and a piston (with same diameter as the internal diameter of the measuring cylinder). The movement of the piston is controlled by a screw thread. The piston is inserted into the base of the measuring cylinder and can be used to extrude 1 mm of soil per turn of the screw. The cut slices of soils are extruded by rotating the screw one turn and the slices were separated from the remaining sample with the use of dental floss thread and a saw blade of 0.1 mm thickness (Figure 28).

The measuring cylinder was bored into the soil by using a rubber hammer and the soil sample of 5 cm thickness was collected, dried and prepared for Gamma spectroscopic measurements (Figure 29). One of the limitations of this fine soil sampler is when the soil sample contains large stones (size greater than the slice thickness) and/or plant roots reaching deep into the soil, a systematic error is introduced during the cutting of slices of desired thickness.

Figure 29: Sampling technique for the depth distribution of ^{7}Be. A: Material needed for sampling, B: Measuring cylinder hammered into the soil, C: collection of the sample with soil surface on the top, D: cut slices are put into small vessels for drying.

To avoid these errors a site survey was done at the three chosen reference sites. Reference site 1 was chosen to take the soil cores as it contained mainly bare soil with shallow rooted grass. Initially the core was cut into 10 slices (0-2 cm) in 2 mm depth increments. First measurements of these samples with gamma spectrometry showed that ⁷Be is confined to the upper 6 mm. Following this, much finer soil cutting with 1 mm depth increments was performed to investigate the distribution in more detail.

4.7 Sample processing

A high degree of standardization was pursued during this study, starting with the sampling technique to the Gamma spectrometry measurements on the samples.

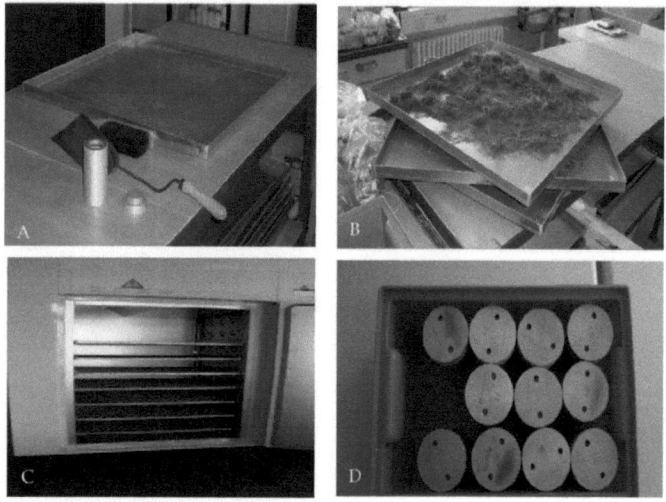

Figure 30: Sample processing for the total inventory of ⁷Be. A: Aluminium tray used for sample drying. B: Sample spread on the aluminium tray. C: Oven used for drying the sample at 105°C. D: Sample geometry used for measurement.

4.7.1 Air drying, grinding, annihilation and sieving for total inventory

The procedure followed to prepare the samples for a total inventory measurement is described here. As a first step the samples were air and oven dried. They were put on aluminium sheets for drying at room temperature and later heated in an oven at 105°C for 22 hours. The sample mass was determined before and after the drying procedure. After drying, the samples were manually disaggregated and passed through an oven to burn the organic material present at 400°C. Afterwards the soil sample was sieved to < 2 mm to remove stones and gravel. The soil mass after sieving was measured again. Finally, the sieved sample was put in aluminium bottles and transported to the gamma spectroscopy lab for measurement (Figure 30).

4.7.2 Fine depth soil sample processing

The procedure for the depth distribution measurements instead was as follows. The soil slices of 1 mm or 2 mm thickness were oven-dried at 80°C for 18 hours. The dried samples were burned if any organic matter was present and then pulverized with pestle and mortar and their dry densities were determined. These samples were then filled in test tubes and transported to the gamma spectrometry laboratory for measurement (Figure 31).

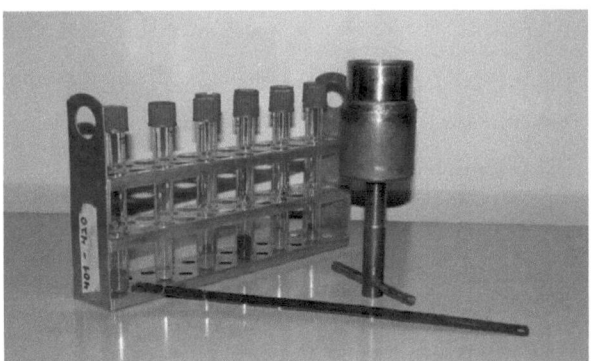

Figure 31: Sample geometry used for the measurement of the ^{7}Be depth distribution at the reference sites.

4.8 Sample analysis

The ^{7}Be concentrations of the soil samples were determined with gamma spectroscopy using high-resolution intrinsic germanium detectors. All samples were analysed at the gamma spectroscopy laboratory of Bundesamt für Strahlenschutz (BfS), Berlin. It is a standard laboratory for the measurement of natural radioactivity and is certified by IAEA through the worldwide open proficiency test performed in the years 2009 and 2010 [106, 107]. Four high-purity Germanium detectors were used for the measurement of ^{7}Be (Figure 32). Their specifications are given in Table 10. Efficiency calibrations of the detectors were performed with the measurement geometry used for different environmental samples during this study.

The efficiency calibration was performed with single-peak sources such as ^{54}Mn, ^{85}Sr, ^{137}Cs, ^{241}Am and ^{109}Cd. In addition ^{57}Co, ^{60}Co and ^{207}Bi sources were also used. The calibration geometry for the efficiency calibration purpose was a 225 ml aluminium bottle with 1 cm; 4 cm and 8 cm fill heights.

The energy calibration was performed for the selected measuring geometry using a standard gamma source known as "Ra-226 Topf" supplied by the Physikalisch-Technische Bundesanstalt (PTB, Braunschweig, Germany). The "Ra-226 Topf" was developed for the naturally occurring radionuclide ^{226}Ra, which is found in all environmental samples. The presence of uranium in the environmental samples would interfere, as the 186 keV gamma photon from ^{235}U cannot be distinguished from ^{226}Ra. Thus the 186 keV peak was not considered for energy calibration purposes. Ra-226 is measured from its progeny Pb-214 and Bi-214. This requires that the noble gas isotope Rn-222 is kept locked gastight in the sample long enough for radioactive equilibrium between Ra-226 and short lived Rn-222 to be established. It takes about three weeks before equilibrium is reached and Ra-226 can be measured using the gamma photons of Pb-214 (352 or 295 keV) and Bi-214 (609 keV). The energies of the measured radionuclides in soil samples collected in this study were calibrated to the energies of the naturally occurring radionuclides.

Table 10: Germanium detectors used for the gamma spectroscopic measurements

Detector	Type	Polarity	Efficiency	Manufacturer
1	Coaxial	n-type	40%	ORTEC
2	Coaxial	p-type	35%	ORTEC
3	Coaxial	n-type	40%	ORTEC
4	Ge Well	p-type	40%	CANBERRA

The gamma spectra were analysed with the Genie 2000 software from Canberra Industries, Inc., Meriden, CT, USA. Beryllium-7 was measured at 477.6 keV as shown in Figure 33. Counting time of 25000-80000 seconds was used to measure ^{7}Be for total inventory (250 ml aluminium bottles) and 80000-560000 seconds for the depth distribution measurements (test tubes).

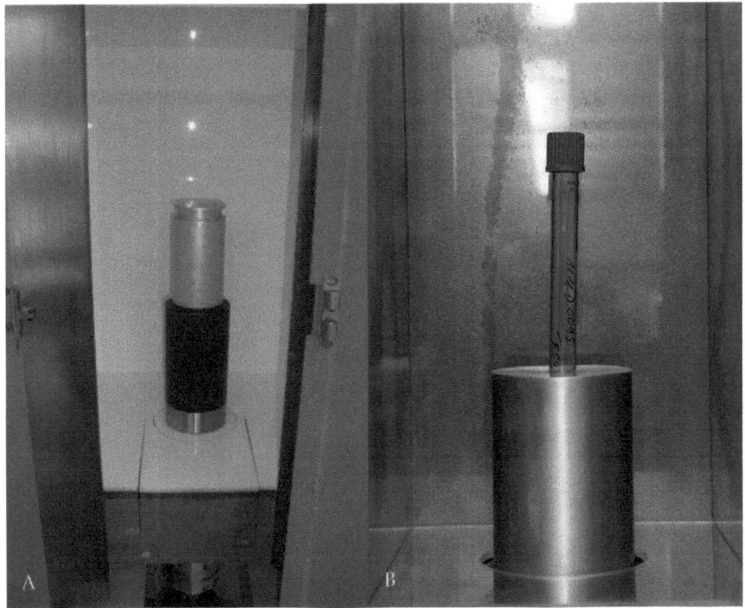

Figure 32: High purity Germanium detectors used to measure ^{7}Be at Bundesamt für Strahlenschutz, Berlin, Germany. A: Coaxial detector used for aluminium bottle geometry, B: Ge Well detector for test tube geometry.

Samples for the depth of 0-1 mm were measured for 80000 sec, as the ^{7}Be is present predominantly on the surface soil layer. Longer measurement times were required for samples below 1 mm, as the areal activity of ^{7}Be below 1 mm decreased sharply leading to less ^{7}Be atoms in the deeper layers compared to the surface layer. Detection limits obtained with the aluminium bottle for soil samples of 300 grams dry mass ranged between 3 and 5 Bq/kg^{-1}. Detection limit for a test tube sample of 2-3 grams measured for 160000 seconds was 10 Bqkg^{-1} (For a probability of 0.05 false positive and false negative errors are, $k_{1-\alpha} = 3$ and $k_{1-\beta} = 1.645$ [108]). Beryllium-7 was assumed to be absent from lower soil core layers when concentration values of two consecutive depth increments were indistinguishable from zero. Measured ^{7}Be activities of the soil samples were corrected for radioactive decay to the time of the sampling with use of Genie 2000 software.

^{7}Be concentrations in the soil samples were measured as fast as possible after the sampling because otherwise they would have been affected by the radioactive decay.

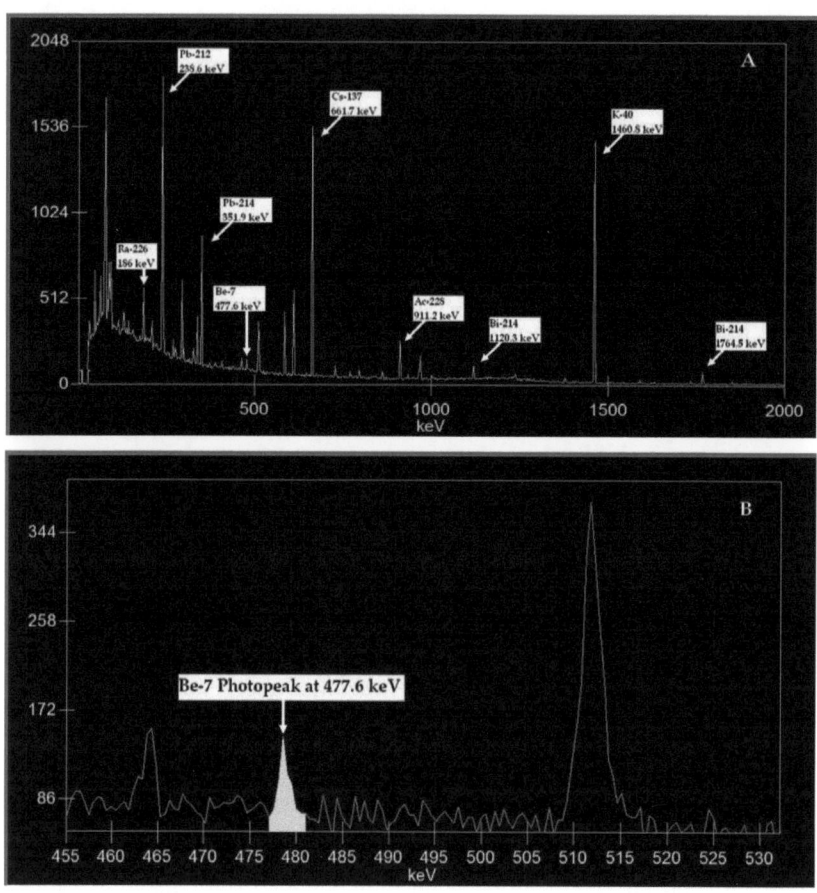

Figure 33: A: Gamma spectrum of soil; B: ^{7}Be photopeak shown at 477.6 keV

4.9 Summary

To explore the potential of ^{7}Be as a short-term soil erosion tracer and to use the radionuclide to calculate erosion rates during multiple rainfall events, a systematic and rigorous sampling approach was planned and executed at the study site in Müncheberg, Germany. The analysis was performed at the gamma spectroscopy laboratory at BfS. The efficiency and energy calibration for all Hp-Ge gamma detectors was conducted with standard sources and measurement geometries of the soil sample. The sampling technique used here was tested frequently, so as to minimize the sampling errors. The fine increment cylinder developed during this study can be used for bare soils or soils with shallow rooted grass over them. Its use is limited when cutting samples containing stones or /and deep grass roots.

5 RESULTS AND DISCUSSION

5.1 Introduction

The first part of this chapter discusses the experimental findings on the depth distributions and total inventories of ^{7}Be at the reference sites. In the second part, erosion rates estimated for steady state and non-steady state conditions are presented and discussed.

5.2 ^{7}Be at the reference site at Müncheberg

The depth distributions and total inventories of ^{7}Be at the study plot in Müncheberg were examined over a period of 19 months. The sampling campaigns were conducted in accordance with the occurrence of rainfall events. In addition, the samples were also collected during dry-periods. The mass of soil collected at the bottom was used to estimate the erosion rate. Samples were taken from the soil in the tin barrier and V-channel and measured with gamma spectroscopy for ^{7}Be activity concentration (Bq kg^{-1}). To establish the depth distribution of ^{7}Be at the reference sites cylindrical cores were collected during the months of May 2011, June 2011, September 2011, October 2011 and November 2011.

5.2.1 Depth distributions of ^{7}Be at the study site

The vertical distribution of ^{7}Be at the reference sites where no erosion or deposition occurs was established from the sample at the reference site 1. Reference site 1 contained mostly soil with short grass. Reference sites 2 and 3 showed a combination of long grass and soil. Because of the limiting ability of the fine increment sampler to cut long grass, reference site 1 was chosen for establishing the depth distribution. Beryllium-7 depth distribution profiles measured at reference site 1 in dry and wet months are presented in Figure 34 and 35. They show that the ^{7}Be profiles have a decreasing concentration with depth. This decline of concentration with depth is exponential in form and is similar to the profiles reported from other studies [18, 19, 21, 33]. Beryllium-7 was detected at the study site up to a depth of 6 mm. Below 6 mm, even with long measurement times; no ^{7}Be activity (Bq kg^{-1}) was detected. After deposition, ⁷Be is sorbed to grain surfaces. This process depends on grain size [30]. Grain size dependencies have been observed for ¹³⁷Cs [112, 113]. At the study site in the topmost soil layer, which is dark in colour and mostly consists of organic material and clay minerals, ⁷Be is quickly adsorbed, leading to high concentration in the surface layer.

During the dry period of the year, the concentration of ⁷Be is higher at the topmost layer. Due to absence of water as the main diffusing agent, the ⁷Be hardly penetrates into the soil (Figure 34).

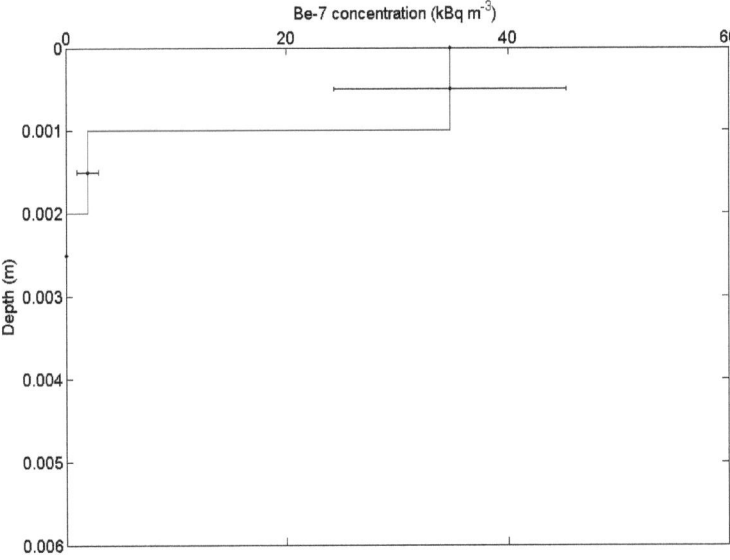

Figure 34: Depth distribution of ⁷Be in the dry period (May 2011). Horizontal error bar is the statistical uncertainty of the measurement.

The measurements confirm the basic process that dominates in the migration of the fallout ⁷Be into the soil column is diffusion. Water is the main carrier of diffusing ⁷Be²⁺ ions.

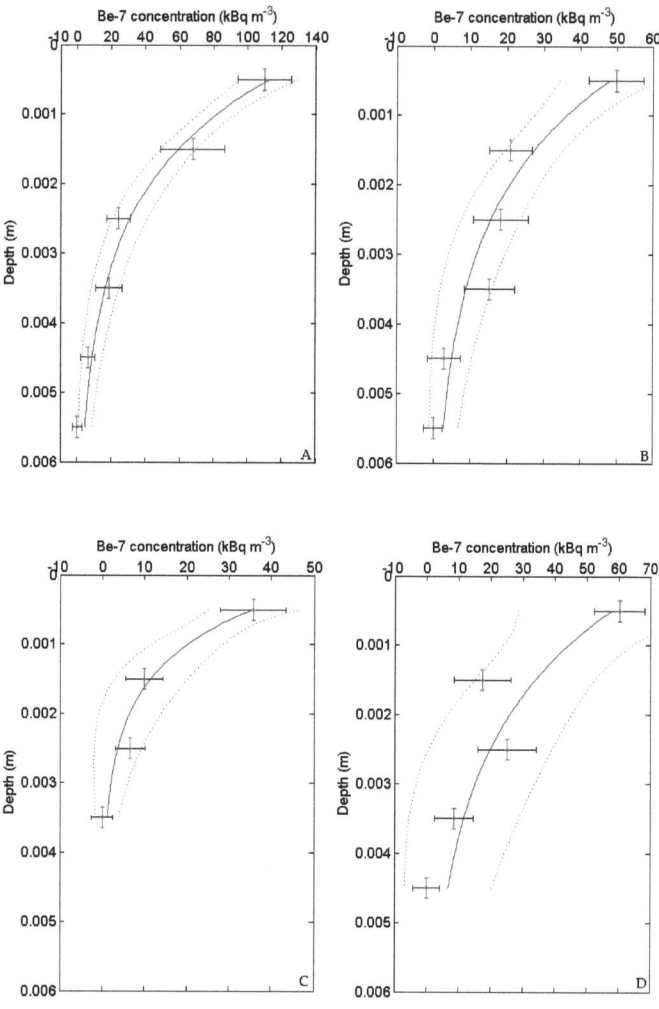

Figure 35: Be-7 concentrations (kBq m^{-3}) with depth for wet months A: June, B: September, C: November and D: October. Blue curve is the statistical fit of the diffusion model (Equation (8)) to the measured data. The dotted curves are the confidence intervals of the fit. Horizontal error bars are the statistical uncertainties of the measurement and the vertical error bars represent the uncertainties over depth.

Beryllium-7 when deposited with the rainwater sorbed to the soil matter and with the solvable phase is subjected to diffusion with the soil solution. Because of sorption at the soil surface and due to its short half-life, $^7Be^{2+}$ ions cannot migrate deeper into soil.

Figure 36 shows the cumulative distribution of ^7Be with depth. It is observed from all depth distribution samples, that 80% of the total concentration of ^7Be in soils is present between the surface (z=0) and 3 mm depth. Below 3 mm the concentrations decrease rapidly. The removal of the upper 1 mm would result in a reduction of 70% of ^7Be at any erosion point.

The effective diffusion coefficient D and the input flux J were estimated by fitting equation (8) i.e. $C(z, t \to \infty) = \frac{I_0}{\sqrt{D\lambda}} e^{-\sqrt{\frac{\lambda}{D}}z}$ with the depth distribution data for ^7Be in the soil at reference site 1 and are presented in Table 11. The fitting procedure is explained in detail in Appendix H. The estimated parameters given in Table 11 were compared with the relaxation mass depth and C_0 by equating the diffusion model (equation (8)) with the empirical profile discussed elsewhere [9, 82].

$$C(z) = C_0 e^{-\frac{z}{h_0}} \quad \text{(Empirical profile)}$$

$$C(z) = \frac{I_0}{\sqrt{D\lambda}} e^{-\sqrt{\frac{\lambda}{D}}z} \quad \text{(Diffusion model (equation (8)))}$$

Comparing the two techniques given above $\therefore C_0 = \frac{I_0}{\sqrt{D\lambda}}$ and $\frac{1}{h_0} = \sqrt{\frac{\lambda}{D}}$, where $C(z)$ (Bq kg^{-1}) is the ^7Be activity at a mass depth z (kg m^{-2}), C_0 is the concentration of ^7Be at the surface (z=0), and h_0 (kg m^{-2}) is the relaxation mass depth which is an empirical description of the depth to which $C = C_0 \frac{1}{e}$. Table 11 shows the diffusion coefficients estimated by using the diffusion model at the reference site for different periods of the year including their uncertainties (Appendix G). The penetration depths h_0 are the depths where the concentration of ^7Be reduces to 1/e times that at the surface. The estimated penetration depths were between 1.4 and 1.8 mm. Similar penetration depths were reported in a previous study for alluvial bare soils [30].

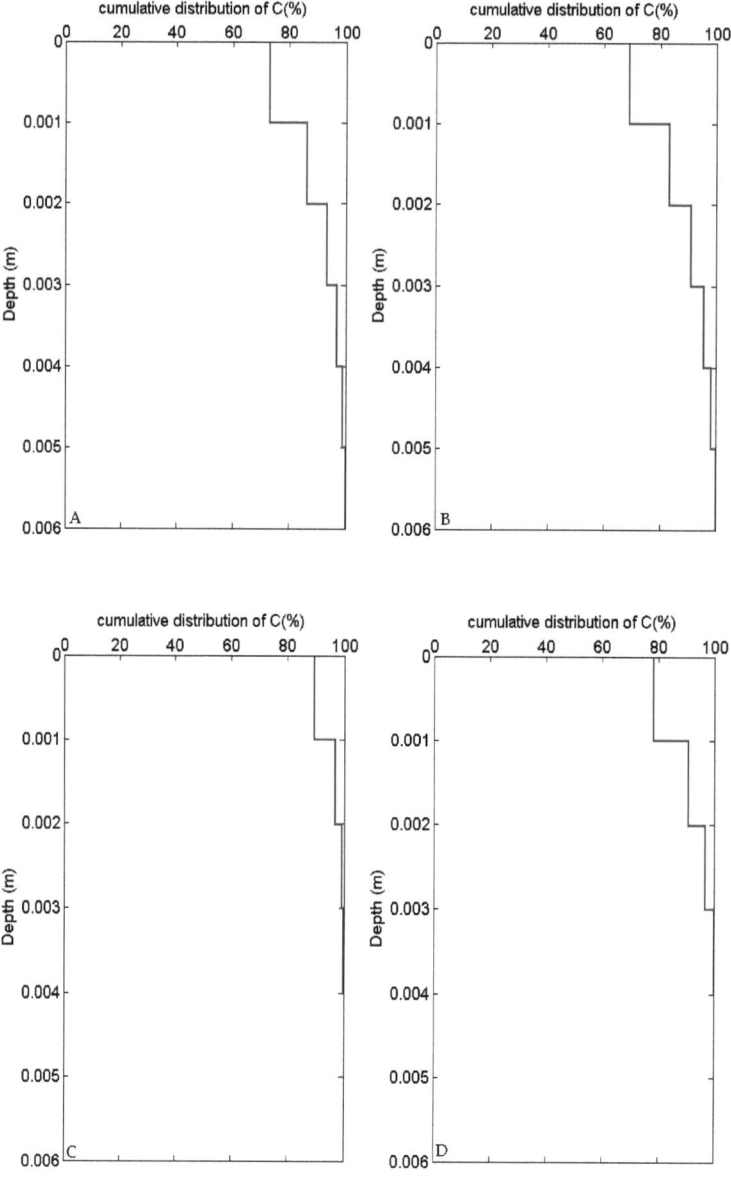

Figure 36: Cumulative depth distribution of ^{7}Be at reference site in Müncheberg for the months June (A), September (B), November (C) and October (D).

Table 11: Values of fit parameters for ^{7}Be at reference site 1 in Müncheberg

Month	Diffusion constant $(D \pm \sigma_D) \times 10^{-13}$ $(m^2\ s^{-1})$	Input Flux $(I_0 \pm \sigma_{I0}) \times 10^{-5}$ $(Bq\ m^{-2}\ s^{-1})$	Relaxation Depth $(h_0 \pm \sigma_{h0})$ (mm)	Surface concentration $(C_0 \pm \sigma_{C0})$ $(kBq\ m^{-3})$
Sept 2010	2.73 ± 0.65	3.43 ± 0.40	1.40 ± 0.68	170 ± 55
May 2011*	NA	NA	NA	NA
June 2011	3.65 ± 0.75	3.64 ± 0.23	1.60 ± 0.70	155 ± 19
Sept 2011	4.75 ± 1.30	1.70 ± 0.20	1.80 ± 1.00	64 ± 15
Oct 2011	2.95 ± 0.90	1.74 ± 0.34	1.40 ± 0.70	83 ± 31
Nov 2011	3.73 ± 0.72	0.90 ± 0.20	1.60 ± 1.00	37 ± 14

*During May 2011 dry periods were documented and diffusion did not take place.

For depth distribution data in May 2011 only two data points were available. Thus, the fitting procedure was meaningless for this profile. May was the driest month in the year 2011. Therefore, the maximum concentration of ^{7}Be in the month of May was found to be at the surface layer. The mean of diffusion coefficient D weighted with the inverse of its uncertainty was calculated and the statistics is presented in Table 12 and it was used in all further calculations of erosion rates.

Table 12: Statistical summary for parameters D and h_0

Parameters	Number of samples (n)	Arithmetic Mean AM ± SD	Weighted mean WM ± WSD
Diffusion coefficient $D\ (m^2\ s^{-1}) \times 10^{-13}$	5	3.56 ± 0.79	3.30 ± 0.33
Penetration depth h_0 (mm)	5	1.56 ± 0.17	1.52 ± 0.34

AM = Arithmetic mean, SD= Standard Deviation, WM= Weighted Mean, WSD= Weighted Standard Deviation

Depth distributions of ^{7}Be were studied in different soils types in other studies [18, 19, 30, 94, 109]. The diffusion model developed in this study was also fitted to these published data and the effective diffusion coefficient and the penetration depths were estimated. The results were compared to the values estimated by the authors during the respective studies. The data is presented in Table 13.

Table 13: Estimated diffusion coefficients and penetration depths for different soils

Sr.No.	Soil type	Source	D (m² s⁻¹) ×10⁻¹³	h₀(Est)* (mm)	h₀ (Lit)** (mm)
1	Alluvial bare	[30]	6.0 ± 1.0	1.9 ± 0.8	1.6
2	Heilu	[109]	17 ± 4	3.3 ± 1.6	NA
3	Very fine clay	[19]	15 ± 1	3.2 ± 0.9	3.6
4	Ultisoils	[19]	17 ± 8	3.4 ± 1.3	2.3
5	Inceptisols	[94]	4.9 ± 0.8	1.8 ± 0.7	NA
6	Ultisoils	[18]	2.7 ± 0.3	1.3 ± 0.4	1.2

*h₀(Est) is the estimated penetration depth by the diffusion model and **h₀(Lit) is the available value of penetration depth from the author of the article. Uncertainties on h₀(Lit) values were not provided by any author.

The data used for estimating the parameters D and h₀ was reconstructed from the research article. Therefore the estimated values have an uncertainty of up to 50%. The diffusion coefficients estimated during this study are the first findings of the study of ⁷Be diffusion in soil. For different soil types as shown in Table 13 the penetration depths, h₀(Est) and h₀(Lit) show similar results. The values of D in Table 13 vary between 2.7×10^{-13} and 17×10^{-13} m² s⁻¹. This indicates that diffusion of ⁷Be in soils does not only depend on the soil type, but is related to the physico-chemical properties of radionuclide transport in soils.

5.2.2 Grain size characteristics of suspended sediments

The particle size distributions for the sediments collected after the erosion event in June 2011 are presented in Figure 37. Figure 37A shows that sediments collected at the V-channel contained 78% clay, 2% silt and 20% sand. Sediments collected at the tin barriers contained 95% sand, 5% silt and 0% clay (Figure 37B). The activity of ⁷Be in the sample collected after the erosion event in June 2011 from the V-channel was 314 ± 20 Bq kg⁻¹ and that from the barrier was 14 ± 1 Bq kg⁻¹ respectively. The specific activities of ⁷Be are proportional to the clay-sized fraction, which is also noted in the other studies [30].

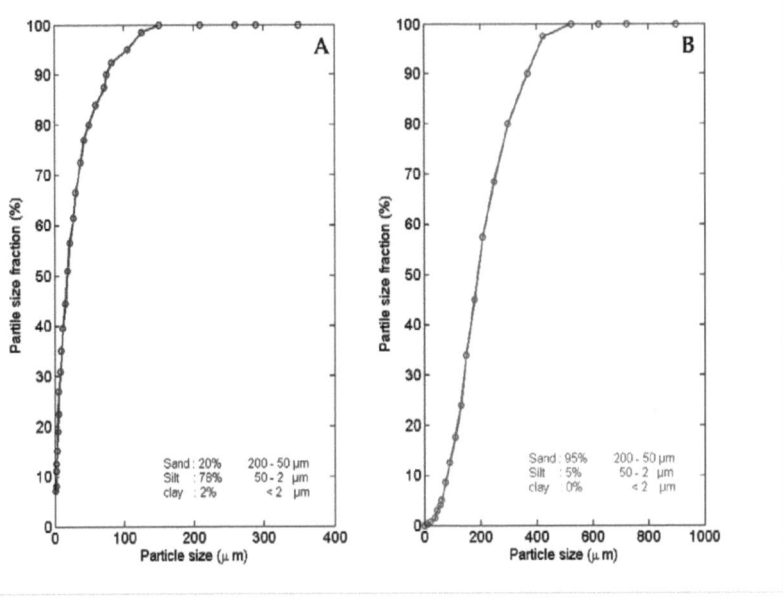

Figure 37: Particle size distribution of suspended sediments discharged during a rainfall event in June 2011 for (A) V-channel and (B) Barrier.

5.2.3 Soil inventories of ^{7}Be at the reference sites

For each rainfall event a weighted mean was calculated for the 3 reference sites RF1, RF2 and RF3 respectively (Table 15). The total inventories of ^{7}Be at the reference sites for the period between March 2010 and October 2011 are presented in Figure 38. The ^{7}Be inventories were measured from March – November in the year 2010 and January - October in the year 2011. In the month of December 2010 the study area was covered with snow and therefore no samples were collected. During the months of July 2010, August 2010, June 2011 and September 2011 the reference sites were sampled twice, as multiple heavy rainfall events occurred and erosion rates were estimated for those situations. The time series of the total inventory data in Figure 38 (top) shows that the ^{7}Be has a peak during the summer (August-November) in both

93

years. Lower inventories were found during the spring and early summer (March-June).

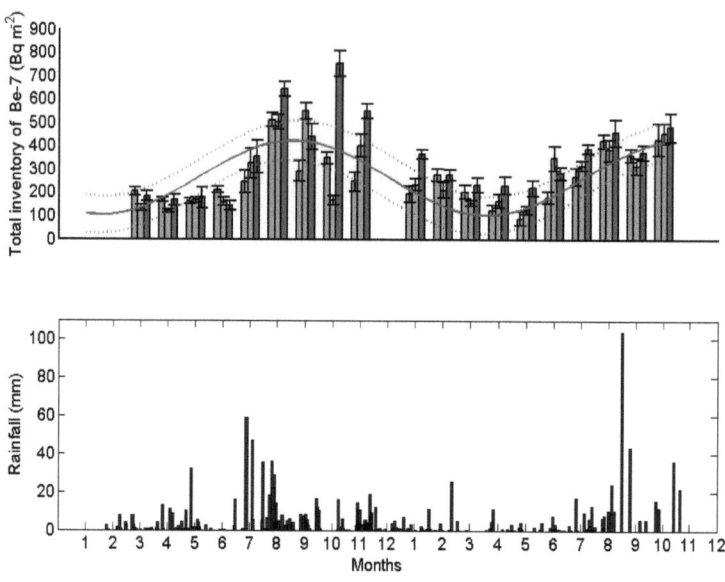

Figure 38: Time series of total inventories of ^{7}Be at the reference sites and Rainfall data for 24 months. In the top plot each group of bar graphs with the measurement uncertainties represents the 3 reference inventories measured in each month with pink, violet and green colours for Reference sites 1, 2 and 3 respectively. The solid red curve is the sinusoidal fit to the data and the dotted lines indicate the 95% confidence bounds to the fit.

The Rainfall data in Figure 38 (bottom) shows that rainfall frequencies are correlated with the increase in ^{7}Be inventories. We applied a two sample Kolmogorov-Smirnov test to check for the identical population between distribution of total inventories of ^{7}Be in 2010 and 2011 respectively. The null hypothesis used for testing the data sets for α = 20% was, H$_0$: Samples have the different continuous distribution. The statistical analysis show that the hypotheses was rejected for α = 20%, P = 0.3911 and Kolmogorov-Smirnov statistic D_{crit} = 0.27. There were no major differences seen between the two distributions. To visualize the difference between two datasets, plot of the two empirical cumulative distribution functions F1(x) (Blue curve) and F2(x) (Red curve) representing ^{7}Be inventories in the years 2010 and 2011 are shown in Figure 39.

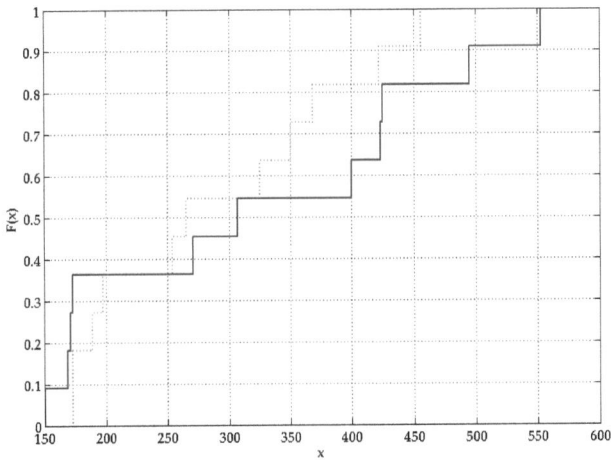

Figure 39: Empirical cumulative distribution plots for the ^{7}Be inventories. Red curve: data set for the year 2010 and Blue curve: data set for the year 2011.

The variability of ^{7}Be at the reference sites was tested using the statistic given in Table 15. Table 15 shows that the variability of ^{7}Be throughout the study period was 10-35%. Considering the periodic behaviour observed at the study site a sinusoidal function $A(t) = \alpha \sin(\omega t - \phi) + \beta$ was fitted to the ^{7}Be total inventories for the period of 2 years. Vertical shift β, Phase shifts Φ, and amplitudes α estimated for the fit are, 160 ± 32 Bq m^{-2}, 0.9 ± 0.2, 264 ± 24 Bq m^{-2}.

There was an exception in the month of October in the year 2010 where the coefficient of variation was 70%. The reason for this was that the samples at RF1 and

RF2 in October were collected from the same points, which were sampled 15 days before. Thus a layer of soil containing ^{7}Be was lost at these sampling points due to the previous sampling campaign. Due to this systematic sampling error the value in October was not used in any further calculation of erosion rates.

Table 14: Total inventories of ^{7}Be for different soil conditions

Month	RF1 (Bq m^{-2}) (bare soil + short grass)	RF2 (Bq m^{-2}) (grass + soil)	RF3 (Bq m^{-2}) (soil + small plants)
Mar-10	204 ± 17	132 ± 17	184 ± 21
Apr-10	169 ± 10	119 ± 10	165 ± 24
May-10	163 ± 12	166 ± 13	177 ± 46
Jun-10	211 ± 15	160 ± 18	143 ± 21
Jul-10	244 ± 17	322 ± 16	315 ± 15
Aug-10	454 ± 21	491 ± 26	542 ± 31
Sep-10	293 ± 46	549 ± 35	441 ± 53
Oct-10	349 ± 26	165 ± 10	754 ± 10
Nov-10	247 ± 41	402 ± 49	551 ± 31
Jan-11	193 ± 36	234 ± 25	367 ± 20
Feb-11	276 ± 29	211 ± 34	275 ± 24
Mar-11	203 ± 31	156 ± 11	233 ± 30
Apr-11	125 ± 23	163 ± 29	230 ± 38
May-11	90 ± 30	126 ± 25	220 ± 20
Jun-11	195 ± 21	267 ± 22	389 ± 21
Jul-11	270 ± 21	316 ± 22	389 ± 21
Aug-11	421 ± 22	371 ± 32	346 ± 22
Sep-11	371 ± 26	374 ± 22	395 ± 27
Oct-11	317 ± 25	360 ± 38	388 ± 56

Reference sites chosen for the sampling exhibited different physical characteristics. Table 14 shows the total inventories of ^{7}Be for the three reference sites RF1, RF2 and RF3. The total inventories of ^{7}Be at all the reference sites show similar behaviour of the radionuclide with some exceptions arising from sampling errors. Atmospheric fluxes given in Table 15 were used for estimation of erosion rates at the study plot.

Table 15: Monthly reference inventories (A_{REF}) of ^{7}Be at Müncheberg

Month-Year	Number of samples	AM (Bq m^{-2})	SD (Bq m^{-2})	WM (Bq m^{-2})	WSD (Bq m^{-2})	SE (Bq m^{-2})	Min (Bq m^{-2})	Max (Bq m^{-2})	CV (%)	σ_{CV} (%)	Atmospheric Flux* (J ± σJ) ×10^{-5} (Bq m^{-2} s^{-1})
Mar-10	3	173	37	165	10	21	132	204	21	9	2.6 ± 0.5
Apr-10	3	151	28	142	6	16	119	169	18	8	2.3 ± 0.4
May-10	3	169	8	165	8	4	163	177	21	8	2.5 ± 0.1
Jun-10	3	171	35	179	10	20	143	211	17	7	2.6 ± 0.5
Jul-10	6	289	51	296	10	21	243	354	18	5	4.4 ± 0.8
Aug-10	6	489	88	491	13	36	392	643	18	5	7.4 ± 1.3
Sep-10	3	495	77	452	25	54	293	549	30	12	7.5 ± 1.2
Oct-10	3	423	301	264	15	174	165	754	71	30	6.4 ± 4.5
Nov-10	3	400	152	434	22	88	247	551	38	16	6.0 ± 2.2
Jan-11	3	265	91	296	14	52	193	367	34	14	4.0 ± 1.3
Feb-11	3	254	37	260	16	22	211	276	15	6	3.8 ± 0.6
Mar-11	3	197	39	170	10	22	156	233	20	8	3.0 ± 0.6
Apr-11	3	173	53	156	16	31	125	230	31	13	2.6 ± 0.8
May-11	3	173	66	135	14	47	90	220	40	15	2.6 ± 1.0
Jun-11	6	230	73	222	14	30	167	349	31	9	3.5 ± 1.1
Jul-11	3	325	60	344	14	35	270	389	18	8	4.9 ± 0.9
Aug-11	6	422	37	381	14	21	328	458	12	3	6.4 ± 0.6
Sep-11	6	404	63	380	15	26	315	482	16	5	6.1 ± 0.9
Oct-11	3	355	36	336	19	21	317	388	10	4	5.4 ± 0.5

AM: Arithmetic Mean, SD: Standard Deviation, SE: Standard Error, Min and Max are the minimum are maximum values in a data set, CV: Coefficient of variation $= \dfrac{SD}{AM} \cdot 100\%$, σ_{CV}: Uncertainty on CV, WM: Weighted Mean, WSD: Weighted standard Deviation, *Atmospheric flux estimated using equation 11

5.3 ⁷Be measurements for soil erosion assessment at the study plot

Erosion rates were estimated for the rainfall events for which soil was collected at the tin barriers and V-channels. 12 events were considered to be short-term erosion events at the study plot during the measuring period. Out of 12 events at the tilled plot, for one event soil redistribution within the plot was observed and no soil was collected at the barrier.

Table 16: Erosion events and plot characteristics at Müncheberg

Months	Year	Observed erosion processes	Canopy cover	Rainfall erosivity EI30 (N m⁻²)	Rainfall intensity I30 (mm h⁻¹)
April	2010	Surface	10%	1	5
May	2010	Splash	100%	11	28
July-1	2010	Surface + Rill	20%	19	28
July-2	2010	Rill	50%	19	37
August-1	2010	Rill	100%	3	17
August-2	2010	Interrill + Rill	100%	26	50
June	2011	Surface	5%	3	22
July	2011	Rill	50%	8	29
August-1	2011	Interrill + Rill	100%	10	12
August-2	2011	Splash	100%	4	22
Sept-1	2011	Rill	100%	1	7
Sept-2	2011	Interrill + Rill	100%	28	59

The quantities related to erosion processes given in Table 16 were measured by the research station situated at the study site. Observed erosion processes were distinguished on the basis of material collected at the tin barrier and its properties. Both EI30 and I30 in Table 16 were estimated with the rainfall and soil simulation model provided by the research institute ZALF [116]. Rainfall erosivity (EI30) is the product of storm energy and maximum rainfall intensity (I30) during a storm. As EI30 increases the erosivity increases, as the kinetic energy of raindrops and the intensity of raindrop impact on the soil surface is increased [110]. The percentage of canopy represents the density of vegetation after ploughing. A canopy cover of 100% is assumed when plants covering the plot attain the maximum height of 1.5 m.

5.3.1 Temporal changes of ⁷Be activities at the study plot

Soil erosion from hill-slopes involves a complex sequence of processes that change during the course of an erosion event. Soil at the study site is transported by overland flow from the top of the slope at 5m to the bottom of the slope where the flow rates were decreased and deposition occurred.

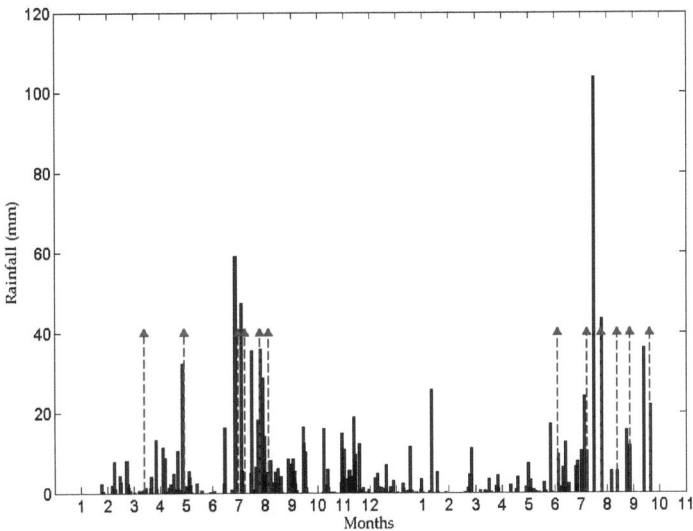

Figure 40: Daily precipitation events for the study period from March 2010- September 2011. Violet arrows indicate the erosion events measured at the study plot.

No severe rainfall events were recorded before April 2010 at the study plot. Therefore it was reasonable to begin the study period for erosion measurement using ⁷Be areal activity across the study plot in April 2010 (Figure 40). Heavy rainfall events with higher intensities (I30 (mm hr⁻¹)) in the months of July, August and September in 2010 and 2011 initiated runoff at both tilled and no-till plots. Soil was removed from the slope and suspended as sediment in the runoff during heavy rainfall events and was collected in the tin barrier; finer soil from the removed material was deposited in the V-channel which was built behind the barrier.

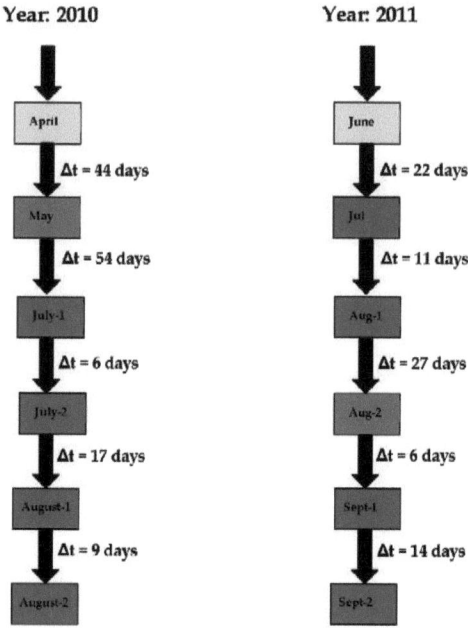

Figure 41: Twelve erosion events occurred during the study period from 2010-2011. Δt represents the time between the erosion events. Colours indicate the erosion processes. (Red: Rill/ interrill erosion, Yellow: Surface erosion, Green: Splash erosion)

The research station at the study site recorded that the rainfall events with intensities greater than 15 mm h^{-1} and lasting for more than 20 minutes caused surface erosion. If rainfall events lasted for several hours, rill erosion was seen at the plot. The material deposited during rill erosion is transported by rill and surface erosion combined. The flow rates during rill erosion are much higher and consequently the carrying capacity of the flow is larger. Figure 42, 43 and 44 shows that the sediments originating from the interrill or surface erosion processes are enriched in clay compared with the composition of the parent soil material. Twelve erosion events occurred in two years were separated by short time periods Δt. Figure 41 shows the erosion processes and the associated time period between the events.

5.3.2 ^{7}Be measurements in the soil at tin barrier and V-channel

The soil collected in the barrier and in the V-channel at the tilled plots was measured for ^{7}Be activities. At the onset of surface erosion during the months of June and July in 2010-11, at the tilled plot, the collected sediments in the tin barriers were associated with relatively large ^{7}Be activity values as shown in Figure 42 and 43.

Erosion events occurred immediately after the plot was tilled. The effect of ploughing is the loosening of the soil, increasing its erodibility. The initial surface erosion process with heavy runoff eroded the uppermost layer of soil. As the darker material eroded from the uppermost layer was transported with the surface flow into the V-channel, which is placed a few meters behind the tin barrier. The heavier sand particles and associated clay and silt were deposited at the tin barrier. As the raindrops also fell on the tin barriers, some of the clay minerals from the sediment in the barriers were also transported to the V-channel. The soil collected in the V-channel was measured separately for ^{7}Be activities. The mass of soil collected at the V-channel was then added to the mass of soil in the tin barrier.

Four heavy rainfall events in the year 2010, as shown in the Figure 41, were separated by 6 days, 17 days and 9 days respectively. It is seen from the depth distribution in Figure 35 that ^{7}Be is mostly concentrated at the uppermost soil layer. As the rainstorms progressed during these four events sediment from the lower soil depths was collected at the barrier as testified by the lower ^{7}Be activities (Figure 45 A). The rills developed along the slope of the plot were observed mainly at 35-50 m on the slope length (Figure 44A).

In 2011 six heavy rainfall events were recorded. The rainfall events were separated by short time intervals of 22 days, 11 days, 27 days, 6 days and 14 days respectively (Figure 41). The intensity and duration of rainfall events in 2011 was higher than in

2010, thus more sand was collected on the barriers with lower ⁷Be activities (Figure 45 B).

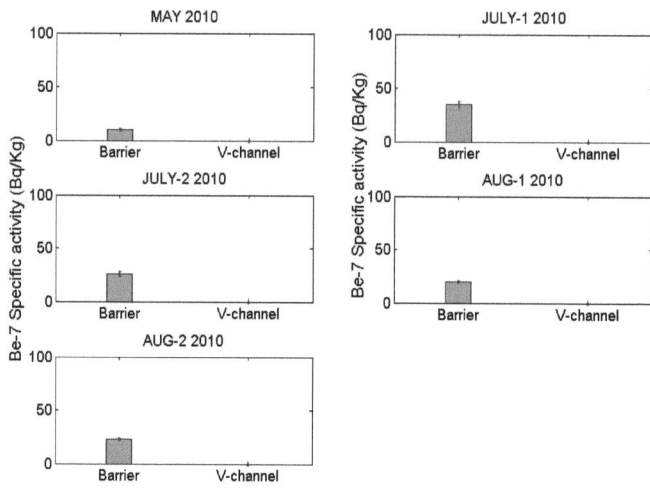

Figure 42: ⁷Be activities in soils collected in tin barriers and V-channels in 2010

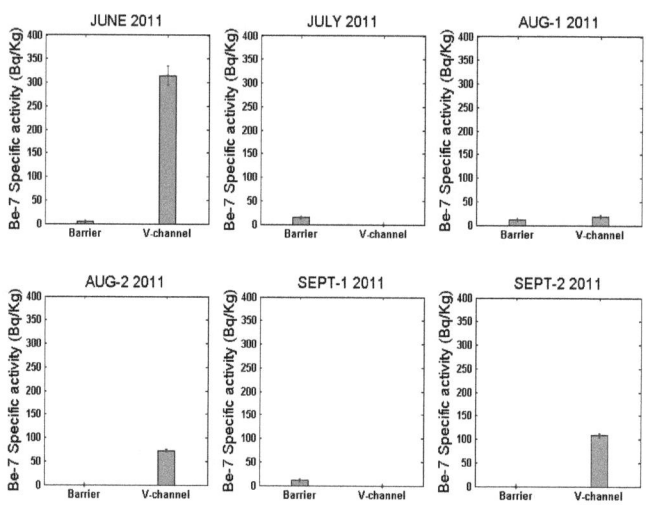

Figure 43: ⁷Be activities in soils collected in tin barriers and V-channels in 2011

Figure 44: Erosion processes observed at the tilled and no-till plots. A: Rill erosion at the tilled plot; B: Surface erosion at the no-till plot

At the no-till plot the main processes observed were surface erosion and pre-rill formation (Figure 44B). Some soil was collected in the barrier at the no-till plot during the months of May, July and August in 2010 and June, July and September in 2011. There was no soil collected in the V-channel at the no-till plot during any erosion event. Beryllium-7 activities in the soil collected in the barrier at no-till plot where high (Table 19). Figure 46 show that the soil at the no-till plot during the study period was mainly redistributed within the plot by means of surface runoff denoting superficial erosion. The mass of soil collected on the barrier was approximately ten times lower than at the tilled plot (Table 19).

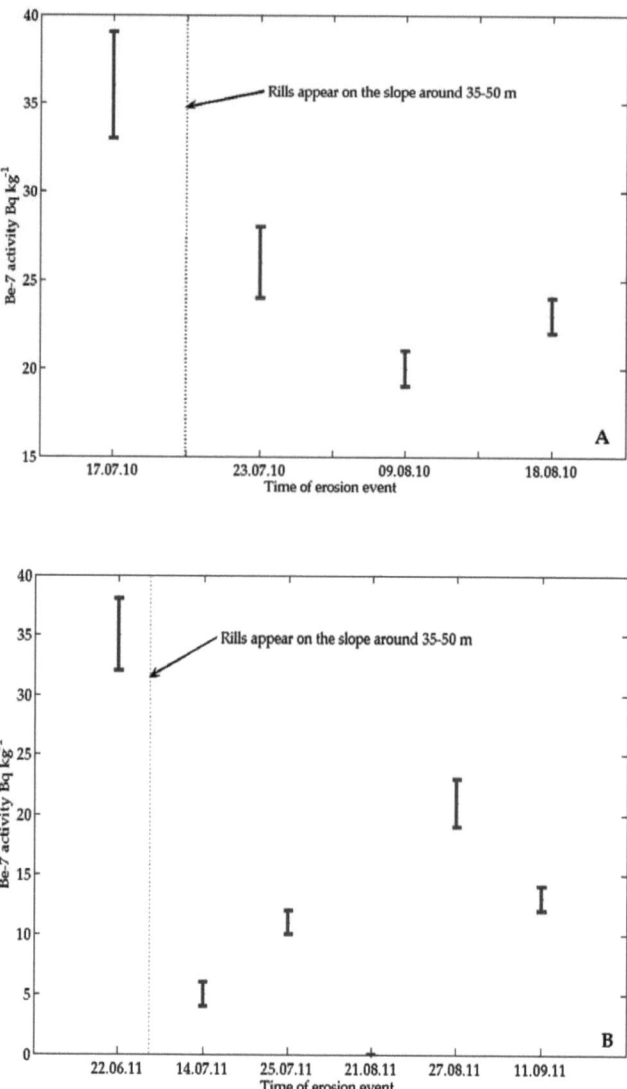

Figure 45: ⁷Be activity of the suspended sediment (barrier + V-channel) samples eroded from the tilled plot for rainfall events in A: 2010, B:2011. The uncertainties indicate the gamma spectrometry measurement precision at 95% level of confidence.

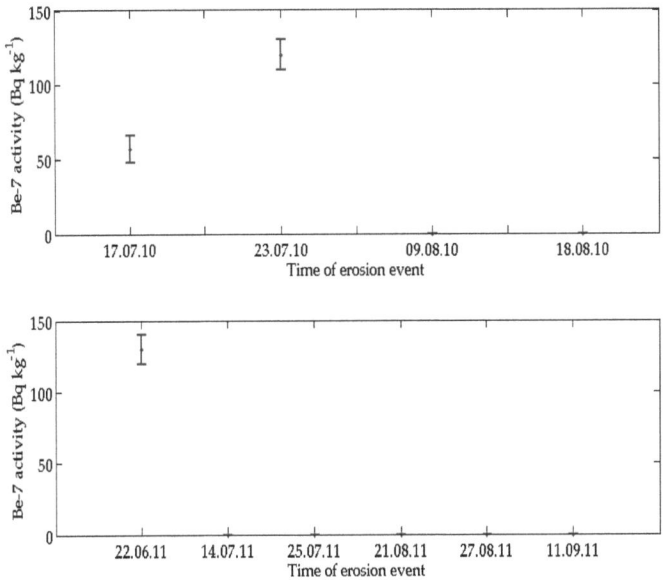

Figure 46: ⁷Be activity of the suspended sediment (barrier + V-channel) samples eroded from the no-till plot for rainfall events in Top: 2010, Bottom: 2011. The uncertainties indicate the gamma spectrometry measurement precision at 95% level of confidence.

5.3.3 Estimation of soil redistribution at the study plot

Erosion events occurred only during the period between June-September, when the atmosphere is instable due to temperature differences, which leads to the formation of clouds, leading to intense rainfall events triggering water erosion. The magnitude of erosion depends on the rainfall intensity, slope length; soil infiltration capacity; slope elevation and canopy cover [111]. At the study area in Müncheberg half of the plot was tilled in the month of June and October. Due to ploughing the soil was homogenized; loosened within upper 25-30 cm and made susceptible to water erosion.

Table 17: Soil redistribution documented for the study site for 12 rainfall events during the years 2010 and 2011, based on ^{7}Be measurements at the tilled plot. The soil redistribution estimates are compared with the physical soil measurements at the study plot.

Year	Month	Erosion (-)/deposition(+) rates at the measurement points along the slope (kg m^{-2})					Collected soil on the barrier (kg)	^{7}Be activity concentration (Bq kg^{-1})	
		5m	20m	35m	50m	53m		Barrier	V-channel
2010	April	< 0.03	< - 0.006	+ 1.5 ± 0.6	- 0.7 ± 0.6	< 0.02	-	-	-
	May	+ 0.7 ± 0.5	+ 0.8 ± 0.5	- 1.0 ± 0.5	+ 1.5 ± 0.6	+ 1.4 ± 0.6	4	10 ± 1	-
	Jul-1	< 0.07	< 0.03	- 1.1 ± 0.5	- 0.9 ± 0.7	< - 0.07	20	35 ± 3	55 ± 3
	Jul-2	- 0.3 ± 0.5	< - 0.001	+ 0.3 ± 0.5	+ 0.6 ± 1	< - 0.003	24	26 ± 2	-
	Aug-1	< - 0.05	- 0.2 ± 0.5	+ 1.4 ± 0.7	+ 1.5 ± 0.5	- 1.8 ± 0.9	4	20 ± 1	-
	Aug-2	- 0.6 ± 0.5	- 1.4 ± 0.9	- 1.8 ± 0.5	- 2.3 ± 0.5	+ 0.7 ± 0.6	145	23 ± 1	24 ± 1
2011	June	- 0.7 ± 0.9	- 0.5 ± 1.0	< - 0.04	- 1.2 ± 0.5	- 0.4 ± 0.6	16	14 ± 2	314 ± 20
	July	+ 1.4 ± 0.5	+ 1.5 ± 0.5	+ 0.5 ± 0.3	+ 2.6 ± 0.8	- 0.9 ± 0.9	170	5 ± 1	19 ± 2
	Aug-1	- 2.1 ± 0.8	- 2.4 ± 0.9	- 4.7 ± 0.4	- 1.5 ± 0.9	+ 1.8 ± 0.7	25	11 ± 1	73 ± 2
	Aug-2	- 0.7 ± 0.5	- 1.2 ± 1.2	+ 0.2 ± 0.6	- 2.2 ± 1.1	- 1.3 ± 0.5	4	-	-
	Sep-1	- 0.5 ± 0.5	< - 0.07	- 0.7 ± 0.8	+ 2.4 ± 0.8	+ 0.8 ± 0.3	13	21 ± 2	-
	Sep-2	- 1.1 ± 1.1	- 0.4 ± 0.8	Rill	Rill	- 5.0 ± 0.8	161	12 ± 1	109 ± 5

Table 18: Sediment delivery ratios based on ⁷Be measurements for the tilled plot at the study site for heavy rainfall events in 2010-11.

Year	Month	Fraction of total eroded area (%)[2]	Fraction of total deposition area (%)[2]	Sediment Delivery Ratio SDR (%)[1]
2010	April	28	28	-
	May	28	72	1
	Jul-1	57	0	3
	Jul-2	9	56	25
	Aug-1	35	56	1
	Aug-2	94	6	7
2011	June	100	0	2
	July	6	94	58
	Aug-1	94	6	1
	Aug-2	72	28	-
	Sep-1	37	34	4
	Sep-2	100	0	8

1) Sediment Delivery Ratio,
$$SDR = \frac{\text{Sediment collected at the barrier/total area of the plot (kg m}^{-2})}{\text{Gross erosion rate calculated using }^7Be \text{ (kg m}^{-2})} \times 100$$

2) Erosion/deposition area fractions were calculated by comparing the total area of the plot and erosion/deposition rates given in the Table 17

The erosion events occurred in the months after ploughing. No-till plot showed comparably lower erosion susceptibility. Rainfall intensities measured at Müncheberg ranged between 7 mm h^{-1} and 59 mm h^{-1}. Water erosion normally occurs at rainfall intensities higher than 7.5 mm h^{-1} and is aggravated for intensities more than 15 mm h^{-1} [110]. During the study period, as shown in Table 16, the measured rainfall intensities (I30) and the rainfall erosivities were sufficient for triggering water erosion. Erosion of soil by water could be seen in different forms at the study site. When the rainfall intensity and rainfall erosivity exceeds the infiltration capacity, surface erosion was observed. Two situations during the study period (May 2010 and August 2011) show splash erosion at the study plot. The study site was ploughed in the months of July in 2010 and June in 2011 at the tilled plot. No-till plot was only harvested in both years. Table 16 shows that 30 days after the ploughing event the vegetation cover had grown up to 100%. The vegetation cover intercepted the incident rainfall at the study plot. The rainwater intercepted by plants falls on the soil surface and becomes part of the surface flow. If the distance of the fall height of drops is high (in case of long plants) drops causes splash erosion.

The magnitude (kg m^{-2}) and pattern of soil redistribution associated with the period of heavy rainfall events occurring in the years 2010 and 2011 were estimated for 12 discrete erosion events using the ^{7}Be flux given in Table 16 and shown by Figure 40 under following two conditions:

1. **Steady state:** The reference inventories were compared with the total inventories of ^{7}Be at the study plot and eroded depths were calculated using equation (20).
2. **Non-steady state:** Simulated ^{7}Be concentrations were compared with the measured inventories at the study plot and eroded depths were calculated using equation (22).

The steady state approach was applied for only one event in April 2010, as before this event no significant rainfall was documented by the weather station at the study plot and ^{7}Be was assumed to have uniformly distributed. For the remaining 11 events the time between the erosion events was short and not sufficient for ^{7}Be concentration to achieve steady state. The plot was ploughed for two events where ^{7}Be concentration was mixed within 25-30 cm. Thus for rainfall events separated by short time intervals and the events occurring after ploughing non steady state approach was used to calculate erosion rates. The uncertainties on the erosion rates were estimated by propagating the measurement uncertainties using the Gauss error propagation method. Uncertainty estimates for the simulated inventory of ^{7}Be were calculated separately using a pseudo Monte Carlo technique (Appendix G2).

The magnitudes of the soil redistribution rate along the slope at the tilled and non-tilled plot calculated using equations (20) and (22) are presented in Tables 17, 18, 19 and 20 respectively. Erosion is shown as negative values and deposition as positive values. The results indicate that the patterns of erosion and deposition are distributed along the slope. The erosion rates are calculated from ⁷Be measurements done at 5m, 20m, 35m, 50m and 53m along the slope length. The radioactive measurements of sediments collected in the barrier and V-channel are also presented in the tables for comparison.

5.3.3.1 Erosion rates at the tilled plots

The total mass of soil collected at the tin barriers during the 6 events in 2010 and 6 events in 2011 was 197 kg and 493 kg respectively.

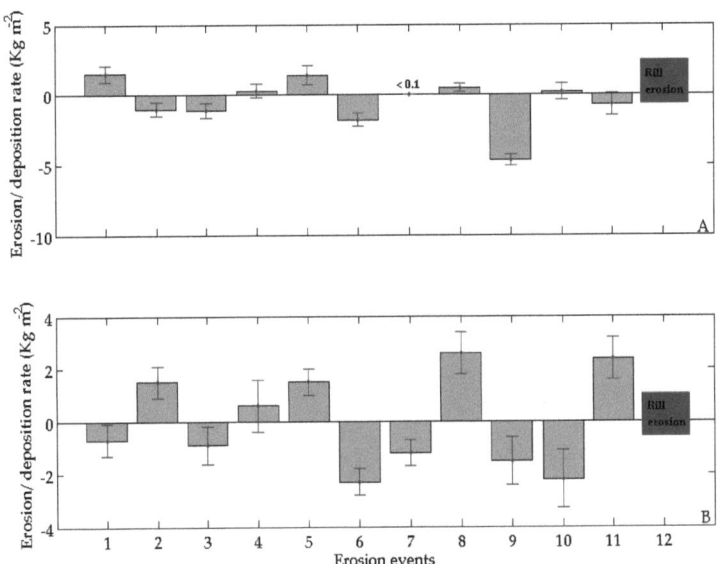

Figure 47: Erosion/deposition rates estimated for 12 erosion events at tilled plot for measurement points A: 35m; B: 50m.

('-': erosion rates, '+': deposition rates)

The erosion rates for the tilled plot in Table 18 show that the erosion rates are estimated between $< - 0.001$ kg m^{-2} and $- 4.7 \pm 0.4$ kg m^{-2} with the largest erosion occurring along the steeper part of the slope from 35-50m. At the top of the slope between 5 m and 20m erosion rates calculated for both the years range from 0 kg m^{-2} to - 2.1 ± 1.5 kg m^{-2}.

Erosion rates estimated at 35m and 50 m on the slope for all erosion events occurred during the study period are presented in Figure 47. During the erosion event in "Sept-2 2011" no ^{7}Be activity was measured in the samples collected at 35m and 50m showing the occurrence of rill erosion. Rills seen on the plot during these erosion events are shown by Figure 47A.

Soil redistribution rates given in Table 17 have high uncertainties. The uncertainty estimates on erosion rates take into account following processes:

1. Differences in total rainfall amounts and the intensities of rainfall during the individual storm events.
2. The percentage of canopy cover present during the heavy rainfall months given in Table 16.
3. The presence of rills, leading to lower ^{7}Be inventories in the soil, which compared with simulated inventories leads to overestimation of erosion rates.

For the erosion events occurring during 2010-11 at the tilled plot the time evolution of the total inventory of ^{7}Be was simulated using equation (14) for a measuring point 50m along the slope length and is presented in Figure 48. After the erosion events occurred in May and Jul-1 in 2010 (as shown in Figure 48 A) and the event occurring in July 2011 the percentage inventory of ^{7}Be reached up to 200%. During these three events soil deposition has occurred, thus the ^{7}Be inventory measured in the sample was higher than the simulated inventory.

For the rainfall events occurring in Jul-1 in 2010 and June in 2011 the ideal conditions for surface erosion with no canopy cover, loose soil and high rainfall intensity were generated. If the eroded material originated from the lower soil depths it cannot contain high ^{7}Be concentrations. At the study site lower ^{7}Be concentrations measured in soil and the higher estimated erosion rates in Tables 17 and 18 shows the rill erosion occurring on the plot.

During the heavy rainfall events in July and August 2010 and August and September 2011, the vegetation cover was up to 100% (plants grown up to 1.5m in height). The intensity of rainfall measured during these months was between 7 and 22 mm h^{-1}. The direct impact of the raindrops was buffered by vegetation. Due to the vegetation, uniform surface erosion was minimized. Some rainfall passed through gaps between the vegetation stripes and part of the intercepted rainfall by vegetation was evaporated from the leaves and the rest fell to the ground and contributed to the surface runoff. As ^{7}Be is deposited with rainfall, after evaporation of water from the trees the ^{7}Be atoms are adsorbed to the tree leaves. Thus, some of the total ^{7}Be flux from atmosphere to ground is reduced. The erosion/deposition rates estimated during these months are shown in Table 17. The higher uncertainties indicated in Table 17 include the effect of vegetation cover.

During the events in August-2 2010 and September-2 2011, a total of 400 kg of soil was collected on the barrier. The rainfall intensities measured during these two events were 50 mm h^{-1} and 59 mm h^{-1} respectively. For events with such high intensities rainfall can pass through the vegetation, even bending it, leading to stronger impacts of raindrops on the soil [117]. This combined with the water falling from the leaves create higher runoff velocities, which lead to surface, rill and interrill erosion [117]. The ^{7}Be activities in the samples from the tin barrier and channel are presented in Table 17.

The sediment delivery ratios in Table 18 show that for the event on "August-2 2010", 7% of the eroded soil was deposited on the tin barrier, while 8% was deposited for the event on "September-2 2011". The reason for low sediment deliveries in spite of large masses of soil in the barriers was that erosion rates during these months were overestimated due to the presence of rills, which contain less ^{7}Be concentrations. Thus the overestimation of erosion rates result in lower sediment deposition ratios. During the " July 2011" event, 170 kg soil was deposited at the tin barriers. The rainfall intensity measured during this event was 29 mm hr^{-1}. The canopy cover during this event was 50%. Due to less canopy cover more rainfall was reached the soil surface and contributed in heavy soil loss. Soil redistribution rates calculated with the ^{7}Be inventories show deposition over the plot (Table 17). This is due to the fact that sampling of soil for the ^{7}Be inventories was conducted at interrill areas. The sediment delivery ratio estimated for this month was 56%, where 6 % of the total area of the plot contributed to the eroded soil (Table 18).

Soil was transported by high intensity rainfall in May and August 2010 and August 2011 at the tilled plot. The plot was fully covered with vegetation during these months (Table 16). The soil collected at the barrier accounted to 4 kg during both the events. The area at the bottom of the slope at around 52-55 m was not covered with vegetation. The rainfall intensities during these events were 28 mm hr^{-1}, 17 mm hr^{-1} and 22 mm hr^{-1} respectively. High intensity rainstorms dislocated soil from the area, which was not covered with vegetation and deposited it into the tin barrier built next to it. The soil redistribution rates estimated for these events show high uncertainties (Figure 49). Large uncertainties on erosion rates in Figure 49 and the amount of soil measured at the barrier indicate that there has not been any significant erosion event occurred during these months at the plot. Field observations made by the Agricultural institute, ZALF also confirm the splash erosion process occurring during these months.

The significant amounts of erosion documented by ^{7}Be measurements and the diffusion model developed in this study demonstrate that, although the presence of vegetation may be effective in reducing moderate rainfall intensities reaching the soil surface, the soil redistribution within the intervening areas can still lead to soil erosion.

During extreme rainfall events the vegetation is not proving effective for shielding the ground from raindrop impact and rill erosion. Soils disturbed by ploughing are especially vulnerable during heavy rainfall events, even if vegetation covered.

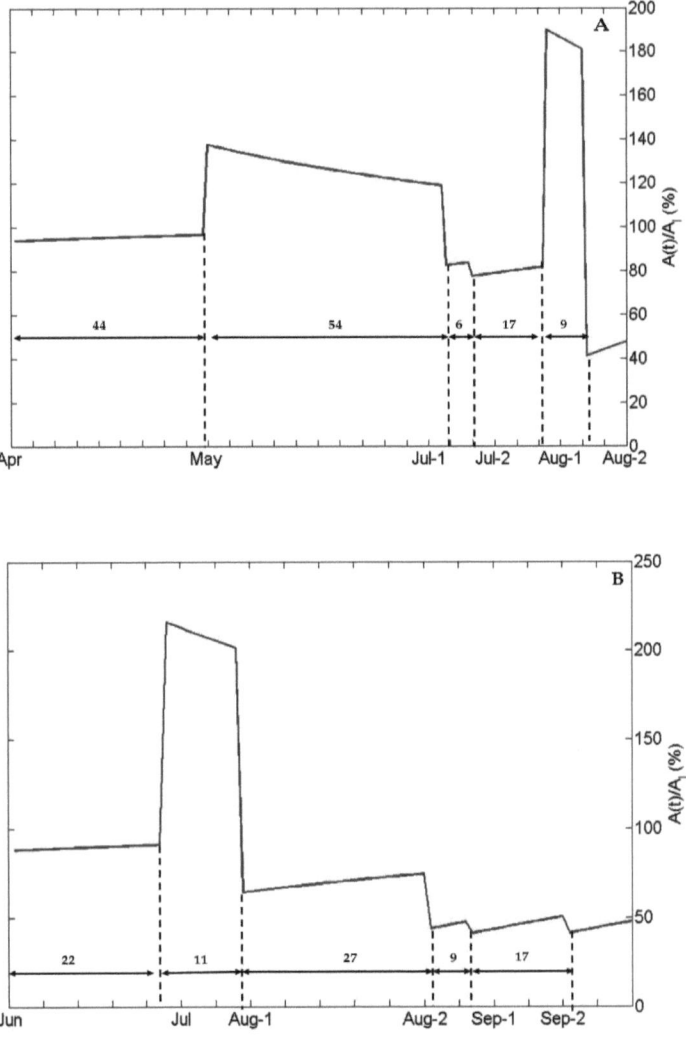

Figure 48: Time evolution of the total inventory of ⁷Be for erosion events occurring in A: 2010; B: 2011, at 50 m along the slope length. The numbers between the erosion events represent the time in days between the events. The curves start from an percentage inventory which is left after an erosion event.

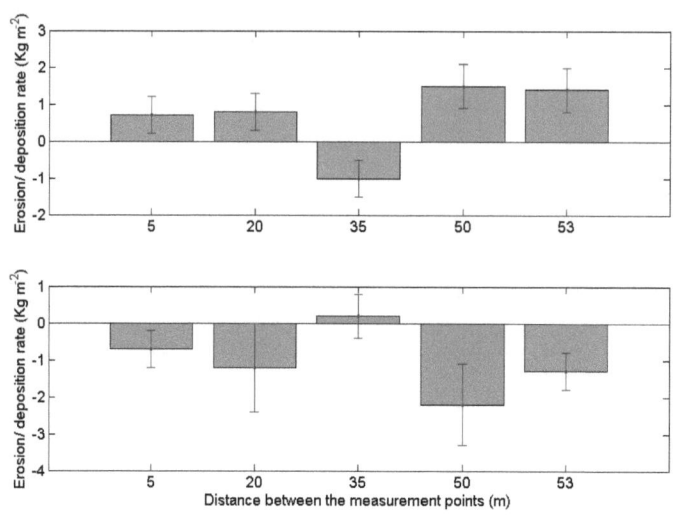

Figure 49: Soil redistribution rates estimated for the tilled plot for the rainfall event in May 2010(Top) and Aug-1 2011(Bottom).

('-': erosion rates, '+': deposition rates)

5.3.3.2 Erosion rates at the no-till plots

The erosion rates estimated for 12 heavy rainfall events using ^7Be inventories are presented in Table 19. Sediment delivery ratios are presented in Table 20. By comparing the erosion rates in Table 17 and 19 we conclude that there was a large difference of magnitude and nature of soil distribution at the no-till plot compared to the tilled plot. For all erosion events in 2010 the sediments collected on the barrier was 9-10 kg, however in 2011 was ~1kg. Amounts of sediments collected in the barriers at the tilled plot were 20-50 times higher than those collected at the no-till plots.

The erosion/deposition rates calculated for no-till plots are estimated with high uncertainties. The plot was covered with vegetation during the heavy rainfall events in July and August 2010 and in July, August and September 2011. The erosion processes at the plot were mainly dominated by surface erosion (Figure 46). If rainfall occurred for longer periods pre-rill structures were formed (Figure 2 (bottom) B).

The impact of rainfall during the heavy and extreme events was reduced by the canopy cover. Uniform erosion was seen during the month of June-2011, when harvesting allowed rainfall to hit the soil surface. The amount of sediment dislocation was prevented by the firm soil structure maintained at the plot. Soil eroded was mostly redistributed within the plot.

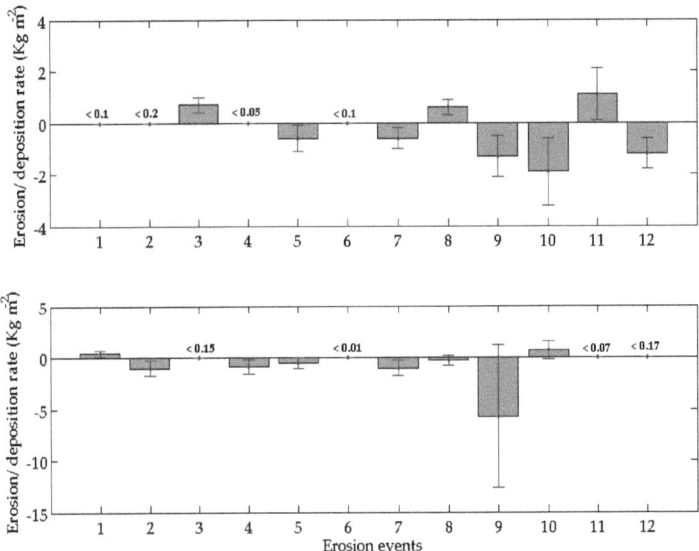

Figure 50: Erosion/deposition rates estimated at no-till plot for 12 erosion events for measurement points 35m (Top plot) and 50m (bottom plot).
('-': erosion rates, '+': deposition rates)

The ^{7}Be tagged sediment transported by runoff was redistributed within the plot, especially at the steepest part of the slope at 35-50 m Erosion estimated at the no-till plot for the measurement points 35 m and 50 m along the slope length are presented in Figure 50. Sediment delivery ratios estimated for both years (Table 20) varied from 0.2 - 1 %. This suggests that the total amount of soil redistribution within the plot was more than the deposition at the barriers. It can be concluded from the available data and the model results that soil erosion at the no-till plot is controlled by no-cultivation practices before eroding months and by canopy cover protecting the soil from erosion by the direct impact of raindrops.

Table 19: Soil redistribution documented for the study site for 12 rainfall events during the years 2010 and 2011, based on ^{7}Be measurements at the no-till plot. The soil redistribution estimates are compared with the physical soil measurements at the study plot.

Year	Month	Erosion (-)/deposition (+) rates at the measurement points along the slope (kg m^{-2})					Soil collected on the barrier (kg)	^{7}Be activity concentration (Bq kg^{-1})	
		5m	20m	35m	50m	53m		Barrier	V-channel
2010	April	<-0.01	+0.8 ± 0.3	<0.05	<-0.02	<-0.04	-	-	-
	May	<0.05	+1.0± 0.3	<0.03	-1.0± 0.7	<0.02	2.5	9 ± 1	-
	Jul-1	+0.5 ± 0.3	+0.8 ± 0.3	+0.7 ± 0.3	<0.01	-0.9 ± 0.3	3	57 ± 9	-
	Jul-2	<0.02	-2.0± 0.5	<0.04	-0.9± 0.7	+1.0 ± 0.8	2	120 ± 10	-
	Aug-1	<0.05	+1.5± 0.3	-0.6 ± 0.5	-0.5 ± 0.5	+1.9 ± 0.3	0.06	-	-
	Aug-2	+0.8 ± 0.3	+1.4 ± 0.2	<0.005	<0.008	-2.3 ± 0.5	1.5	-	-
2011	June	-0.5 ± 0.4	-0.5 ± 0.6	-0.6 ± 0.4	-1.0 ± 0.7	-0.2 ± 0.3	0.1	130 ± 10	-
	July	+0.9 ± 0.3	-0.3 ± 0.4	+0.6 ± 0.3	-0.3 ± 0.5	<0.04	0.3	-	-
	Aug-1	-1.1 ± 0.9	<-0.07	-1.3 ± 0.8	-5.7 ± 6.9	-0.5 ± 0.8	-	-	-
	Aug-2	+0.8 ± 0.5	<-0.06	-1.9 ± 1.3	+0.7 ± 0.9	-2.1 ± 1.4	-	-	-
	Sep-1	<-0.01	-0.7 ± 0.6	+1.1 ± 1.0	<0.09	+1.5± 0.3	-	-	-
	Sep-2	-0.5 ± 0.5	+0.4 ± 0.3	-1.2 ± 0.6	<0.07	-0.8 ± 0.5	0.3	-	-

Table 20: Soil redistribution based on ⁷Be measurements for the no-till plot at the study site for heavy rainfall events in 2010-11.

Year	Month	Fraction of total eroded area (%)[2]	Fraction of total deposition area (%)[2]	Sediment Delivery Ratio SDR (%)[1]
		0	38	
2010	April	28	29	-
	May	6	63	1
	Jul-1	57	6	1
	Jul-2	56	44	0.2
	Aug-1	6	38	0
	Aug-2	100	0	0.2
2011	June	57	37	-
	July	100	0	0.3
	Aug-1	34	37	-
	Aug-2	29	34	-
	Sep-1	100	0	-
	Sep-2			-

2) Sediment Delivery Ratio,

$$SDR = \frac{\text{Sediment collected at the barrier/total area of the plot (kg m}^{-2})}{\text{Gross erosion rate calculated using }^{7}\text{Be (kg m}^{-2})} \times 100$$

2) Erosion/deposition area fractions were calculated by comparing the total area of the plot and erosion/deposition rates given in the Table 19

5.3.4 Comparison of erosion rates calculated by ⁷Be and direct soil measurements from ZALF

Soil budget was calculated for each measurement point at the study plot. For this purpose a cumulative calculation of ⁷Be-tagged soil at each measurement point was done. The plot was divided into 6 areas surrounding the measurement points from A_1 to A_6, which contribute to erosion or deposition. Each measurement point contributes to either erosion or deposition as given in the tables 17 and 19 respectively. The soil influx and outflux at each measurement point was weighted with the corresponding area. If k_1, k_2, k_3, k_4, k_5 are the masses of soil dislocated from each measurement point and deposited on the area situated between them (Figure 51), the soil depositing on the tin barriers is estimated as:

$$M_{EST} = A_6 \cdot k_5$$

Where M_{EST} = Mass of soil depositing on the tin barrier (kg); A_6 = Area between measurement point at 53m and the tin barrier (m²) and k_5 = dislocated soil from point 5 and deposited on area A_6 (kg m⁻²).

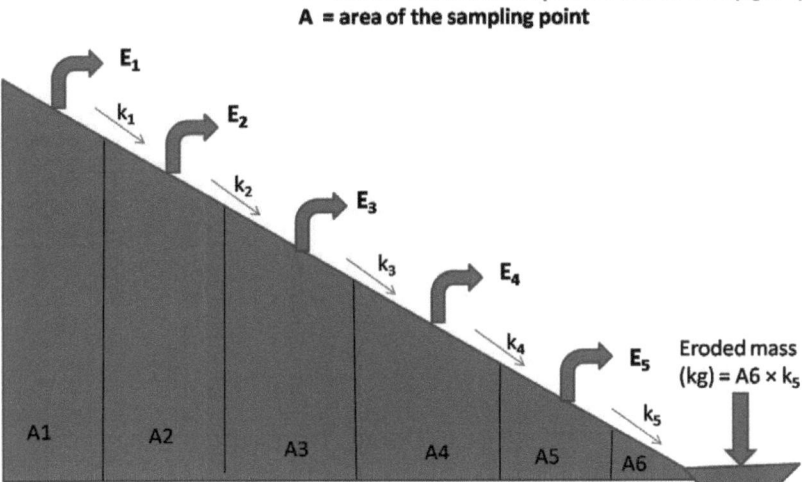

Figure 51: Soil budget calculation at the study plot

The soil deposited at the barrier from the tilled and no-till plot and was estimated as shown in figure 51 and is given as 'M_{EST}' in tables 21 and 22. This estimated soil is compared direct measurements performed by the research institute ZALF and is presented in both the tables as 'M_{ZALF}' The estimated soil with the associated uncertainties at the study plot by ⁷Be measurements shows a good agreement with the physical soil measurements.

Table 21: Comparison of estimated soil using ^{7}Be technique and the direct measurements at the tin barrier at the tilled plot

Months/ Events		Soil budget at the measurement points					Measured soil, M_{ZALF} (kg)	Estimated soil, M_{EST} (kg)
		5m	20m	35m	50m	53m		
April 2010	ERO (-)	-	-	-	- 0.7	- 0.5	-	3 ± 3
	DEP (+)	-	-	+ 1.5	-	+ 0.5		
May 2010	ERO (-)	-	-	- 1	- 1.2	-	-	0 ± 0
	DEP (+)	-	-	+ 1.5	-	+ 0.5		
Jul-1 2010	ERO (-)	-	-	- 1.1	- 1.2	- 0.8	20	20 ± 10
	DEP (+)	-	-	-	+ 0.3	+ 0.8		
Jul-2 2010	ERO (-)	- 0.3	- 0.08	- 0.32	- 0.7	- 0.9	24	22 ± 36
	DEP (+)	-	+ 0.08	+ 0.02	+ 0.1	+ 0.9		
Aug-1 2010	ERO (-)	-	- 0.2	- 1.35	- 1.85	- 0.6	4	14 ± 26
	DEP (+)	-	-	+ 0.05	+ 0.4	+ 1.2		
Aug-2 2010	ERO (-)	- 0.6	- 1.6	- 2.2	- 3	- 1.3	145	122 ± 38
	DEP (+)	-	+ 0.2	+ 0.2	+ 0.06	+ 0.8		
Jun 2011	ERO (-)	- 0.7	- 0.7	- 0.2	- 1.26	- 1.2	16	29 ± 35
	DEP (+)	-	+ 0.2	+ 0.2	+ 0.06	+ 0.8		
Jul 2011	ERO (-)	-	-	-	-	- 0.9	170	21 ± 19
	DEP (+)	+ 1.4	+ 1.5	+ 0.5	+ 2.6	-		
Aug-1 2011	ERO (-)	- 2.1	- 3	- 5.5	- 3.2	- 0.4	25	10 ± 5
	DEP (+)	-	+ 0.6	+ 0.8	+ 1.7	+ 2.2		
Aug-2 2011	ERO (-)	- 0.7	- 1.4	- 0.2	- 2.3	- 2.9	4	68 ± 39
	DEP (+)	-	+ 0.2	+ 0.4	+ 0.1	+ 1.5		
Sep-1 2011	ERO (-)	- 0.5	- 0.2	-	-	-	13	0 ± 0
	DEP (+)	-	+ 0.2	+ 0.5	+ 2.4	+ 0.8		
Sep-2 2011	ERO (-)	- 1.1	- 0.7	- 5.5	- 6.4	- 9	161	213 ± 91
	DEP (+)	-	+ 0.3	+ 0.2	+ 1.7	+ 4.3		

Table 22: Comparison of estimated soil using ^{7}Be technique and the direct measurements at the tin barrier at the no-till plot

Months/ Events		Soil budget at the measurement points					Measured soil, M_{ZALF} (kg)	Estimated soil, M_{EST} (kg)
		5m	20m	35m	50m	53m		
April 2010	ERO (-)	-	-	-	-	-	-	0 ± 0
	DEP (+)	+ 0.7	+ 0.8	-	-	-		
May 2010	ERO (-)	-	-	-	- 1	- 0.7	3	4 ± 3
	DEP (+)	-	1	-	-	+ 0.7		
Jul-1 2010	ERO (-)	-	-	-	-	- 0.9	3	5 ± 2
	DEP (+)	+ 0.5	+ 0.8	+ 0.7	-	-		
Jul-2 2010	ERO (-)	-	- 2	- 0.5	- 1.1	- 0.3	2	1.6 ± 0.7
	DEP (+)	-	-	+ 0.5	+ 0.2	+ 0.7		
Aug-1 2010	ERO (-)	-	-	- 0.6	- 0.7	-	0.06	0 ± 0
	DEP (+)	-	+ 1.5	-	+ 0.2	+ 0.5		
Aug-2 2010	ERO (-)	-	-	-	-	- 2.3	1.5	9 ± 4
	DEP (+)	+ 0.8	+ 1.4	-	-	-		
Jun 2011	ERO (-)	- 0.5	- 0.6	- 0.7	- 1.2	- 1	0.1	6 ± 4
	DEP (+)	-	+ 0.2	+ 0.2	+ 0.2	+ 0.8		
Jul 2011	ERO (-)	- 0.9	-	- 0.08	-	- 0.2	0.3	1.2 ± 1.0
	DEP (+)	-	+ 0.3	-	+ 0.3	+ 0.2		
Aug-1 2011	ERO (-)	-	- 0.3	- 0.08	- 0.4	- 4.1	-	28 ± 19
	DEP (+)	+ 1.1	+ 0.3	+ 1.4	+ 6.1	+ 4.6		
Aug-2 2011	ERO (-)	-	-	-	- 0.1	- 2.7	-	16 ± 17
	DEP (+)	-	-	+ 1.9	+ 0.6	+ 0.07		
Sep-1 2011	ERO (-)	-	- 0.7	-	-	-	-	0 ± 0
	DEP (+)	-	-	+ 0.2	-	+ 1.5		
Sep-2 2011	ERO (-)	- 0.5	-	- 1.2	- 0.4	- 1	0.3	6 ± 9
	DEP (+)	-	+ 0.2	-	+ 0.4	+ 0.3		

The uncertainties on the direct measurements were not available. The comparison shown in the tables 21 and 22 clearly states that the first hand validation of erosion rates at different parts of the slope can be achieved by with ^{7}Be model presented in this work. The higher uncertainties on the erosion rates are partly due to the reason that the diffusion model in study does not include the other physical processes such as infiltration rates, canopy cover, soil roughness etc. Another source of uncertainties is, Beryllium-7 technique estimates erosion rates over the complete plot and direct measurements assume the uniform erosion over the complete study plot. To validate the erosion rates a physical process based model, which can quantify erosion rates over the whole plot and at the same points as the ^{7}Be measurements, must be used.

The comparison shown in the Tables 21 and 22 clearly states that the validation of erosion rates cannot be achieved by comparing the sediment collected in the barriers and ^{7}Be measurements at different parts of the slope alone. To validate the erosion rates a physical process based model, which can quantify erosion rates over the whole plot and at the same points as the ^{7}Be measurements, has to be used. Another approach could be planting erosion pins at the measurement points along the slope, which could provide an estimate of surface lowering or accretion in the immediate vicinity of the pin.

6 CONCLUSIONS AND OUTLOOK

6.1 CONCLUSIONS

Main goal of this study included: Establishing a physical process based model to study the vertical behaviour of ^{7}Be in soils and use this model to quantify erosion rates for single and multiple rainfall events. The results presented in this thesis confirm the potential of ^{7}Be as a tracer in soil erosion investigations for single and multiple storm events. By providing estimates of soil redistribution rates associated with 12 individual erosional events over a period of two years at our study site in Müncheberg, Germany, offers a valuable complement to ^{137}Cs, which has been more widely used in soil erosion studies.

The diffusion model proposed in this study to our knowledge is the first extensive model proposed so far that despite of its many simplifications, adequately represents the exponential distribution of ^{7}Be profiles at disturbed and undisturbed or reference sites. Mathematical simulations performed using diffusion model show the scope of this model to be used in various scenarios such as multiple rainfall events separated by the short time interval Δt and erosion events occurring after the field is ploughed.

The proposed diffusion model has been fitted to more than 15 depth distributions, including the profiles taken at our study site in Müncheberg and the available depth distributions in the literature. The migration parameters of ^{7}Be in soil were calculated by fitting the soil profiles to the diffusion model. In general diffusion coefficients estimated were of the order of $10^{-12} - 10^{-13}$ m^2 s^{-1} for loamy to sandy soil types. Diffusion coefficients estimated for our study site were about 10^{-13} m^2 s^{-1}. Main physical processes, which transport of ^{7}Be in soil is molecular diffusion and radioactive decay decreases its concentration in soils. Migration parameters and measurements confirm that sorption is the main physical process, which confines ^{7}Be concentration to soil surface.

Current erosion estimation methods with ^{7}Be available in the literature for estimating erosion rates for single rainfall event was successfully modified to quantify erosion rates for multiple rainfall events. Estimated erosion rates at tilled and no-till plots were between < 0.001 - 4.7 ± 0.4 kg m^{-2} and 0.3 ± 0.5 kg m^{-2} - 2.0 ± 1.4 kg m^{-2} respectively. The magnitude of erosion rates estimated at the no-till plots was less than that at the tilled plots. Heavy erosion was estimated at the steepest part of the slope between 35m-50m along the slope length. Erosion rates estimated with ^{7}Be technique were successfully used to differentiate between the rill-interrill, splash and surface erosion at the study plot.

A limitation of the Diffusion model proposed in this study is it does not take into account the vegetation cover. Thus during the months when vegetation is present on the field the model, in some cases, overestimates the erosion rates or in some cases shows the occurrence of deposition on the plot.

Ploughing was performed twice a year at the study plot. The ploughing event that occurred before the heavy rainfall period led to the higher soil movement. To avoid this, specifically for the agricultural practices based on slopes one time ploughing method should be practised.

6.2 OUTLOOK

The current approach of erosion rates quantification for multiple rainfall events can be used successfully for discrete rainfall events. But for a situation where the surface lowering (erosion) is continuously occurring then the current approach cannot be used. For this purpose a mathematical model consisting of differential equations approach with a moving boundary should be used.

The validation is required for the erosion quantification technique presented in this thesis. For this purpose a more physical process based models which include the effects of wind, vegetation, water content, solar radiation etc., should be used. Some of the well implemented approaches found in the literature include, Revised Universal Soil Loss Equation (RUSLE), EROSION- 3D and WEPP model.

The plot used in this study contained continues slope. This plot was specially built for the erosion studies. The need to upscale the use of ^{7}Be technique from the small field of few m² to the watershed of few hundred m² will need the application of additional tools (e.g. GIS, Global Positioning Technique (GPS) and Geostatistics), to assist in integrating and interpreting spatial complexity of the landscape.

The sampling technique used in this study for establishing depth distribution has a limitation against soils with deep rooted grass or with stones. More refined technique should be developed for the fine cutting of the soil layers to document the ^{7}Be depth distribution. One approach to tackle this problem is presented in the literature in the form of Fine Soil Increment Corer (FSIC) [113].

Developing an improved understanding of the post fallout behaviour of ^{7}Be in soils and related environments (e.g. canopy interception, preferential adsorption/ desorption mechanisms) is one of the scientific issue that should be dealt in order to refine the ^{7}Be technique.

APPENDIX

Appendix A: 1-D diffusion equation without radioactive decay

A1: In terms of concentration C (z, t)

The 1-D diffusion equation without the sink $S_i = -\lambda C(z,t)$ is given as,

$$\frac{\partial C(z,t)}{\partial t} = -D\frac{\partial^2 C(z,t)}{\partial z^2} \tag{B1}$$

The solution of equation (B1) is presented here in terms of the main mathematical steps.

Taking Laplace transform of equation (B1) w.r.t. t, L_t results in

$$\int_0^\infty \frac{\partial C(z,t)}{\partial t} e^{-st} dt = D\int_0^\infty \frac{\partial^2 C(z,t)}{\partial z^2} e^{-st} dt \tag{B2}$$

The variable in Laplace space corresponding to t is denoted by s. $L_t[C(z,t)]$ is written as $\overline{C} = \int_0^\infty C(z,t)e^{-st} dt$. Using UV integration rule on the LHS of (B2) we get:

$$L_t\left\{\frac{\partial C(z,t)}{\partial t}\right\} = s\overline{C} - C(s,t=0) = s\overline{C} \text{ (as per the initial condition)}$$

$$\therefore s\overline{C} = -D\frac{\partial^2 \overline{C}}{\partial z^2} \tag{B3}$$

The solution of (B3) is given as:

$$\overline{C} = Ae^{-\sqrt{\frac{s}{D}}z} \tag{B4}$$

Here A is the constant of integration which is evaluated by differentiating (B4) w.r.t. z and equating it to (B1) at z=0 we have $A = \dfrac{I_0}{s\cdot\sqrt{s}\cdot\sqrt{D}}$. Thus (B4) can be rewritten as:

$$\overline{C} = \frac{I_0}{s\cdot\sqrt{s}\cdot\sqrt{D}} e^{-\sqrt{\frac{s}{D}}z} \tag{B5}$$

The back-transform or the inverse Laplace transform is straightforward if the transform tables are used. Thus the concentration C (z, t) is evaluated after taking inverse transform of (B5) and is given by (B6).

$$C(z,t) = I_0 \left[\left(\frac{2\sqrt{t}}{\sqrt{\pi D}} e^{-\frac{z^2}{4Dt}} \right) - \frac{z}{D} \text{erfc}\left(\frac{z}{2\sqrt{Dt}} \right) \right] \quad (B6)$$

Test for validity of the solution

Integrating (B6) over the complete soil column i.e. from 0 to ∞ to test the solution (B6), if it represents the boundary and initial conditions used in the differential equation.

$$\int_0^\infty C(z,t)dz = I_0 \int_0^\infty \left(\frac{2\sqrt{t}}{\sqrt{\pi D}} e^{-\frac{z^2}{4Dt}} \right) dz - I_0 \int_0^\infty \frac{z}{D} \text{erfc}\left(\frac{z}{2\sqrt{Dt}} \right) dz$$

$$= 2I_0 t - I_0 t$$

$$\therefore \int_0^\infty C(z,t)dz = I_0 t$$

Here, $\text{erfc}(x) = \frac{e^{-x^2}}{x\sqrt{\pi}}$ (neglecting the higher order terms)

$$\int_0^\infty e^{-x^2} dx = \frac{\sqrt{\pi}}{2}$$

The above test shows that the soil column contains the flux J at all times t which confirms the validity of the solution

A2: Solution of diffusion equation (B1) in terms of flux J_0

Consider J_0 as a constant flux per unit area per time deposited on the surface of the soil column. The flux $J(x, t)$ is given by the Fick's law (B7). Using (B7) in (B1) we can write

$$J(x,t) = -D\left(\frac{\partial C(z,t)}{\partial z} \right) \quad (B7)$$

$$D\frac{\partial^2 J(x,t)}{\partial z^2} = \frac{\partial J}{\partial t} \quad (x>0, t>0) \quad (B8)$$

Equation (B8) is solved with the use of Laplace transform with boundary condition at $x = 0, t > 0$ as $J = J_0$ (constant) [114].

$$J = J_0 \text{erfc}\left(\frac{z}{2\sqrt{Dt}} \right) \quad (B9)$$

Rearranging and integrating (B7) from z=z to z=∞,

$$C(z,t) = \frac{1}{D}\int_z^\infty J \, dz \qquad (B10)$$

Using (B9) in (B10) and using the properties of error function as

$$i^0 \text{erfc}(x) = \text{erfc}(x) \,\, \& \,\, \text{ierfc}(x) = \frac{1}{\sqrt{\pi}} e^{-x^2} - x\text{erfc}(x),$$

$$C(z,t) = J_0\left[\left(\frac{2\sqrt{t}}{\sqrt{\pi D}} e^{-\frac{z^2}{4Dt}}\right) - \frac{z}{D}\text{erfc}\left(\frac{z}{2\sqrt{Dt}}\right)\right] \qquad (B11)$$

The solution can be identified with the well defined solution given elsewhere in the literature for continues flux boundary condition [114].

Appendix B

Part 1: Laplace transforms technique for solving diffusion equation with radioactive decay

Taking Laplace transform of Equation (5) w.r.t. t, L_t on both sides and using the initial condition we get,

$$\therefore (s+\lambda)\overline{C} = D\frac{\partial^2 \overline{C}}{\partial z^2} \qquad (B12)$$

Rearranging (B12) we get the quadratic equation:

$$\frac{\partial^2 \overline{C}}{\partial z^2} - \left(\frac{s+\lambda}{D}\right)\overline{C} = 0 \qquad (B13)$$

the solution of (B13) is given as (Neglecting the divergent term),

$$\overline{C} = Ae^{-\sqrt{\frac{s+\lambda}{D}}z} \qquad (B14)$$

To Find the constant A, Laplace transform of the boundary condition is equated to (B14) for z=0.

$$-D\frac{\partial \overline{C}}{\partial z} = \frac{I_0}{s} \Rightarrow \frac{\partial \overline{C}}{\partial z} = -\frac{I_0}{sD}$$

differentiating (B14) w.r.t z and equating it to the equation above for z=0

$$A\left(-\sqrt{\frac{s+\lambda}{D}}\right) = -\frac{I_0}{sD} \Rightarrow A = \frac{I_0}{s \cdot \sqrt{s+\lambda} \cdot \sqrt{D}} \tag{B15}$$

Using (B15) in (B14) we get the result as shown by equation (8) in chapter 2.

Part 2: 1-D diffusion equation with the pulse boundary condition (5b)

Applying the Laplace transformation corresponding to t, L_t, to equation (5) results in the solution in the Laplace space as shown in Part 1 of Appendix B.

$$\overline{C} = Ae^{-\sqrt{\frac{s+\lambda}{D}}z} \tag{B16}$$

To evaluate constant A, the Laplace transformation of (5b) is taken w.r.t. t. This result in,

$$-D\frac{\partial \overline{C}}{\partial z} = I; \text{ where}\left(L_t\{\delta(t)\}=1\right)$$

$$\Rightarrow -DAe^{-\sqrt{\frac{s+\lambda}{D}}z} = I$$

Using A in (B16) we get,

$$\overline{C} = \frac{I}{\sqrt{D} \cdot \sqrt{s+\lambda}} e^{-\sqrt{\frac{s+\lambda}{D}}z} \tag{B17}$$

Transforming (B17) as $\frac{s+\lambda}{D} = s' \Rightarrow ds = Dds'$, $s = Ds' - \lambda$ and taking inverse Laplace transformation as explained in Appendix C,

$$L_t^{-1}\left\{\frac{e^{-x\sqrt{s}}}{\sqrt{s}}\right\} = \frac{1}{\sqrt{\pi Dt}} e^{-\frac{x^2}{4Dt}} \cdot e^{-\lambda t} \tag{B18}$$

Part 3: Time evolution of total inventory of ^{7}Be after erosion

The time evolution of the concentration of ^{7}Be from time t=0 to t = t' is calculated by integrating (13) from 0 to t'.

$$\int_{t=0}^{t=t'} \frac{dC}{(I_0 - \lambda \cdot C(t))} = \int_{t=0}^{t=t'} dt$$

We have $\int \frac{1}{A+B \cdot X} = \frac{1}{B}\ln(A+B \cdot X) + \text{constant}$

$$\therefore t = \left[-\frac{1}{\lambda}\ln(I_0 - \lambda C(t))\right]_0^{t'}$$

$$\therefore t' = -\frac{1}{\lambda}\ln\left(\frac{(I_0 - \lambda C(t'))}{(I_0 - \lambda C_0)}\right) \quad (\text{As at } t=0\ C=C_0,\ \text{constant}=0) \tag{B19}$$

Rearranging (B19) and using (11) we can write,

$$A(t') = \frac{I_0}{\lambda}\left[1 - \left(e^{-\lambda t} - e^{-\lambda t}\cdot e^{-\sqrt{\frac{\lambda}{D}}\cdot\Delta z}\right)\right] \tag{B20}$$

Part 4: Time evolution of depth distribution of ^7Be after erosion

Applying Laplace transformation to (5) w.r.t. t, L_t we get,

$$L_t\left[\frac{\partial C}{\partial t}\right] = L_t\left[D\frac{\partial^2 C}{\partial z^2}\right] - L_t[\lambda C] \tag{B21}$$

$$L_t\left[\frac{\partial C}{\partial t}\right] = -C(z, t=0) + s\overline{C} \tag{B22}$$

$\therefore$ Using (15), equation (B22) becomes

$$-C_E + s\overline{C} = D\frac{\partial^2 \overline{C}}{\partial z^2} - \lambda \overline{C}$$

$$\therefore \frac{\partial^2 \overline{C}}{\partial z^2} - \left(\frac{s+\lambda}{D}\right) = -\frac{C_E}{D} \tag{B23}$$

The exact solution to equation (B23) is written in two parts,

$$\text{Exact solution} = \text{Complementary Function} + \text{Particular Integral}$$

The complementary function is given by,

$$\text{C.F.} = A\exp\left(-\sqrt{\frac{s+\lambda}{D}}z\right) \tag{B24}$$

Taking Laplace transform of (B24) and equating to equation (B) at z=0 we get,

$$\text{C.F.} = \frac{I_0}{s\sqrt{s+\lambda}\sqrt{D}}\exp\left(-\sqrt{\frac{s+\lambda}{D}}z\right) \tag{B25}$$

The particular integral is found out by the following way,

$$C_E = \frac{C_s \lambda}{\sqrt{D\lambda}}e^{-\sqrt{\frac{\lambda}{D}}(Z+\Delta Z)} \Rightarrow C_E = C_{01}e^{-\sqrt{\frac{\lambda}{D}}z} \tag{B26}$$

Where,

$$C_{01} = \frac{I_0}{\sqrt{D\lambda}} e^{-\sqrt{\frac{\lambda}{D}}\Delta z} = \text{constant}$$

The Particular Integral for equation (B23) can be chosen as,

$$PI = Ke^{-\sqrt{\frac{\lambda}{D}}z} \tag{B27}$$

Where K = constant

To find K differentiating (B27) w.r.t. z twice and using equation (B23) we get,

$$\frac{\lambda}{D} Ke^{-\sqrt{\frac{\lambda}{D}}z} - \left(\frac{s+\lambda}{D}\right) Ke^{-\sqrt{\frac{\lambda}{D}}z} = \frac{-C_0 e^{-\sqrt{\frac{\lambda}{D}}z}}{D} \tag{B28}$$

Rearranging (B28),

$$K = \frac{C_0}{s} \tag{B29}$$

Using (B29) in (B27),

$$PI = \frac{C_0}{s} e^{-\sqrt{\frac{\lambda}{D}}z} \tag{B30}$$

The exact solution to (B23) is given by adding (B26) and (B30),

$$\overline{C} = \frac{I_0}{s\sqrt{s+\lambda}\sqrt{D}} \exp\left(-\sqrt{\frac{s+\lambda}{D}}z\right) + \frac{C_0}{s} \exp\left(-\sqrt{\frac{\lambda}{D}}z\right) \tag{B31}$$

Here $\overline{C}$ is the concentration of ^7Be in the Laplace space. The depth distribution of ^7Be $C(z, t)$ evolved in time t and depth z is obtained by taking an inverse Laplace transformation of (B31) using standard Laplace transformation tables.

$$C(z,t) = \frac{I_0}{2\sqrt{D\lambda}} \left[e^{-\sqrt{\frac{\lambda}{D}}z} \text{erfc}\left(\frac{z}{2\sqrt{Dt}} - \sqrt{\lambda t}\right) - e^{\sqrt{\frac{\lambda}{D}}z} \text{erfc}\left(\frac{z}{2\sqrt{Dt}} + \sqrt{\lambda t}\right) \right] + C_E \tag{B32}$$

Appendix C: Inverse Laplace transformation

The inverse Laplace transform is evaluated by using Cauchy's formula as:

$$C = \frac{I_0}{\sqrt{D}} \int_{\gamma-i\infty}^{\gamma+i\infty} \frac{e^{-z_1\sqrt{s+\lambda}}}{s\sqrt{s+\lambda}} \cdot e^{st} dt; \quad z_1 = \frac{z}{\sqrt{D}} \tag{B33}$$

Where γ is a real number selected so that all singularities of L(s) are to the left of the line $s = \gamma$, i.

Using $s + \lambda = s' \Rightarrow ds = ds'$, $s = s' - \lambda$ & $a^2 = \lambda \Rightarrow a = \sqrt{\lambda}$ in equation (6)

$$C = \frac{I_0 e^{-\lambda t}}{\sqrt{D}} \int_{c-i\infty}^{c+i\infty} \frac{e^{-z_1 \sqrt{s'}}}{(s' - a^2)\sqrt{s'}} \cdot e^{s't} dt \qquad (B34)$$

According to the inverse Laplace transform property,

$$L_t^{-1}\left\{\frac{e^{-x\sqrt{s}}}{(s-a^2)\sqrt{s}}\right\} = \frac{C-D}{2a};$$

$$C = e^{a^2 t - ax} \operatorname{erfc}\left(\frac{x}{2\sqrt{t}} - a\sqrt{t}\right) \&$$

$$D = e^{a^2 t + ax} \operatorname{erfc}\left(\frac{x}{2\sqrt{t}} + a\sqrt{t}\right)$$

Using above properties in (B34), (B31) and (B18) the analytical solution to 1-D diffusion equation is obtained for different initial and boundary conditions.

Appendix D: Crank-Nicholson discretization technique for 1-D diffusion equation

D1: Finite differences approximations [115]

The finite differences method gives an approximate solution for $C(z,t)$ at a finite set of z and t. The discrete z used in this thesis are uniformly spaced in the interval $0 \leq z \leq L$ such that

$$z_i = (i-1)\Delta z, \qquad i = 1, 2, \ldots N$$

N is the total number of spatial nodes including those at the boundary. L is the thickness of the soil column.

The spacing between the z_i is computed as

$$\Delta z = \frac{L}{N-1}$$

Similarly, the discrete t are uniformly spaced in $0 \leq t \leq t_{max}$:

$$t_m = (m-1)\Delta t, \quad m = 1, 2, \ldots M$$

Here M represents the number of time steps and Δt is the size of a time step

$$\Delta t = \frac{t_{max}}{M-1}$$

The solution domain is depicted in Figure 52. Table 22 summarizes the notation used to obtain the approximate solution to (5) and to analyse the result.

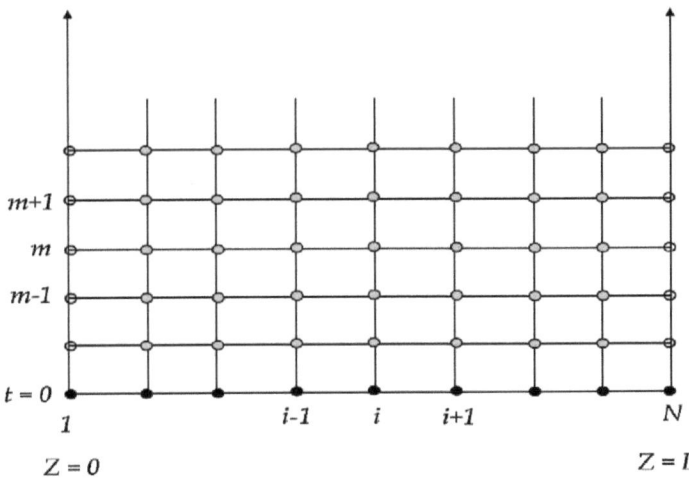

Figure 52: Mesh on semi infinite strip used for solution to the one-dimensional diffusion equation (6). The solid dots indicate the location of the (known) initial values. The open dots indicate the location of the (known) boundary values. The blue circles indicate the position of the interior points where the finite difference approximation is computed.

Table 22: Notation of variables

Symbol	Meaning
C(z,t)	Continuous solution (true solution)
C(z_i,t_m)	Continuous solution evaluated at the mesh points
C_i^m	Approximate numerical solution obtained by solving the finite differences equations.

D2: Crank-Nicolson scheme for 1D diffusion equation for ^7Be

Consider equation (5),

$$\frac{\partial C(z,t)}{\partial t} = -D\frac{\partial}{\partial z}\left(\frac{\partial C(z,t)}{\partial z}\right) - \lambda C(z,t)$$

The right hand side of (5) was discretized by using the average of central differences scheme evaluated at current and previous time steps (Figure 53)

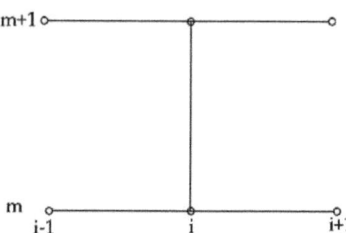

Figure 53: Computational nodes for Crank-Nicolson finite differences technique

The finite differences formulas used to solve (5) numerically are listed in the following way:

1. First order central difference

Writing C_{i+1} and C_{i-1} in terms of Taylor series expansions and solving for $\left(\frac{\partial C}{\partial z}\right)_z$ gives,

$$\left.\frac{\partial C}{\partial z}\right|_{z_i} = \frac{C_{i+1} - C_{i-1}}{2\Delta z} + \sigma(\Delta x^2) \qquad (B35)$$

$\sigma(\Delta z^2)$ is the truncation error. The size of truncation error is kept small as we chose the mesh size Δz. As $\Delta z \ll 1$ the truncation error of central differences scheme goes much faster to zero than the forward differences scheme.

2. Second order central difference

For the higher order derivatives in z the Taylor series expansion is approximated about $C(z_i)$ which gives following result:

$$\left.\frac{\partial^2 C}{\partial z^2}\right|_{z_i} = \frac{C_{i+1} - 2C_i + C_{i-1}}{\Delta z^2} + \sigma(\Delta x^2) \qquad (B36)$$

3. Backward time scheme in spatial dimension

The left hand side of equation (5) is approximated using implicit backward in time scheme. The backward in time scheme is unconditionally stable and has a temporal truncation error of $\sigma(\Delta t^2)$

$$\left.\frac{\partial C}{\partial t}\right|_{t_{m+1}, z_i} = \frac{C_i^m - C_i^{m-1}}{\Delta t} + \sigma(\Delta t^2) \tag{B37}$$

Using equation (B36) and (B37), equation (5) is approximated with

$$\frac{C_i^m - C_i^{m-1}}{\Delta t} = \frac{D}{2}\left[\frac{C_{i-1}^m - 2C_i^m + C_{i+1}^m}{\Delta z^2} + \frac{C_{i-1}^{m-1} - 2C_i^{m-1} + C_{i+1}^{m-1}}{\Delta z^2}\right] - \frac{\lambda}{2}\left[C_i^m + C_i^{m-1}\right] \tag{B38}$$

Rearranging equation (B44),

$$-\frac{D}{2\Delta z^2}C_{i-1}^m + \left(\frac{1}{\Delta t} + \frac{D}{\Delta z^2} + \frac{\lambda}{2}\right)C_i^m - \frac{D}{2\Delta z^2}C_{i+1}^m$$
$$= \frac{D}{2\Delta z^2}C_{i-1}^{m-1} + \left(\frac{1}{\Delta t} - \frac{D}{\Delta z^2} - \frac{\lambda}{2}\right)C_i^{m-1} + \frac{D}{2\Delta z^2}C_{i+1}^{m-1} \tag{B39}$$

Boundary condition discretization

The boundary condition (4a) at $z = 0$ and $z = L$ is discretized using (B39),

At $z=0$ (at the surface) the approximation of (4a) follows,

$$-\frac{D}{\Delta z^2}C_2^m + \left(\frac{1}{\Delta t} + \frac{D}{\Delta z^2} + \frac{\lambda}{2}\right)C_1^m = \left(\frac{1}{\Delta t} - \frac{D}{\Delta z^2} - \frac{\lambda}{2}\right)C_1^{m-1} + \frac{D}{\Delta z^2}C_2^{m-1} + \frac{2I_0}{\Delta z} \tag{BC1}$$

At $z=L$ (at the bottom)

$$\frac{\partial C(z,t)}{\partial z} = 0 \tag{B40}$$

Thus (B46) is approximated as,

$$C_{N+1}^m - C_{N-1}^m + C_{N+1}^{m-1} - C_{N-1}^{m-1} = 0 \tag{BC2}$$

From (B45), BC1 and BC2, the system of equations is solved at each time step. The coefficients of the interior nodes, a_i, b_i, c_i and d_i are defined as:

$$a_i = -\frac{D}{2\Delta z^2}; \; b_i = \frac{1}{\Delta t} + \frac{D}{\Delta z^2} + \frac{\lambda}{2}; \; c_i = -\frac{D}{2\Delta z^2}$$
$$d_i = \frac{D}{2\Delta z^2}C_{i-1}^{m-1} + \left(\frac{1}{\Delta t} - \frac{D}{\Delta z^2} - \frac{\lambda}{2}\right)C_i^{m-1} + \frac{D}{2\Delta z^2}C_{i+1}^{m-1}$$

Equation (B45) can be written as,

$$a_i C_{i-1}^m + b_i C_i^m + c_i C_{i+1}^m = d_i \tag{B41}$$

Equation (B41) was efficiently solved by the standard tridiagonal system method using LU factorization with backward substitution (Appendix E).

Appendix E: Solving a tridiagonal system [115]

The system of equations for Crank-Nicolson scheme can be represented in matrix form as Av =d.

where

$$A = \begin{bmatrix} b_1 & c_1 & 0 & 0 & 0 & 0 \\ a_2 & b_2 & c_2 & 0 & 0 & 0 \\ 0 & a_3 & b_3 & c_3 & 0 & 0 \\ 0 & 0 & a_4 & b_4 & c_4 & 0 \\ 0 & 0 & \ddots & \ddots & \ddots & 0 \\ 0 & 0 & 0 & a_{N-1} & b_{N-1} & c_{N-1} \\ 0 & 0 & 0 & 0 & a_N & b_N \end{bmatrix}, \quad v = \begin{bmatrix} C_1 \\ C_2 \\ C_3 \\ \vdots \\ \vdots \\ C_{N-1} \\ C_N \end{bmatrix}, \quad d = \begin{bmatrix} d_1 \\ d_2 \\ d_3 \\ \vdots \\ \vdots \\ d_{N-1} \\ d_N \end{bmatrix}$$

The system of equations given above is solved by Lower Upper tridiagonal matrix factorization (LU decomposition) and with the backward substitution. Since the coefficient matrix is known to be positive definite (& symmetric in case of diffusion without convection), we can write LU factorization without pivoting.

If coefficient matrix is A, we wish to find Lower and Upper diagonals such that A=LU.

here L and U have the form

$$L = \begin{bmatrix} e_1 & 0 & 0 & 0 & 0 & 0 \\ a_2 & e_2 & 0 & 0 & 0 & 0 \\ 0 & a_3 & e_3 & 0 & 0 & 0 \\ 0 & 0 & a_4 & e_4 & 0 & 0 \\ 0 & 0 & \ddots & \ddots & 0 & 0 \\ 0 & 0 & 0 & a_{N-1} & e_{N-1} & 0 \\ 0 & 0 & 0 & 0 & a_N & e_N \end{bmatrix}, \quad U = \begin{bmatrix} 1 & f_1 & 0 & 0 & 0 & 0 \\ 0 & 1 & f_2 & 0 & 0 & 0 \\ 0 & 0 & 1 & f_3 & 0 & 0 \\ 0 & 0 & 0 & 1 & f_4 & 0 \\ 0 & 0 & 0 & \ddots & \ddots & 0 \\ 0 & 0 & 0 & 0 & 1 & f_{N-1} \\ 0 & 0 & 0 & 0 & 0 & 1 \end{bmatrix}$$

Evaluating each non-zero term in the product LU and setting it equal to the corresponding entry in A, gives

$$e_1 = b_1$$
$$e_1 f_1 = c_1$$

$$a_2 = a_2$$
$$a_2 f_1 + e_2 = b_2$$
$$e_2 f_2 = c_2$$
$$\vdots \quad \vdots$$
$$a_n = a_n$$
$$a_n f_{n-1} + e_n = b_n$$

Solving for the unknowns e_i and f_i gives:

Since A=LU, the system Av=d is equivalent to the (LU)v =d. Introducing the w vector defined as w =Uv, the system of equations become Lw=d. Since L is the lower triangular matrix, w can be easily obtained by solving Lw = d. Thus with w known, v is computed by solving Uv = w. Thus once L and U have been found, the v vector is computed in two step process.

$$\text{Solve Lw = d}$$
$$\text{Solve Uv = d}$$

Since L is the lower triangular, the first solve is forward substitution. Since U is upper triangular, the second solve is backward substitution.

Appendix F: MATLAB code to solve Diffusion equation by Crank-Nicolson scheme

```
%*********************************************************************%
% --- Define constants and initial condition---%
clc;
clear;
```

```
L    = 0.02;                        % length of domain in x direction in meters
J    = 0.0000254791*24*60*60;       % Bq m⁻² day⁻¹
L1   = (0.693/53.2);                % Decay constant in day⁻¹
tmax = 25;                          % End time
nx   = 2000;                        % Number of nodes in x direction
nt   = 1500;                        % Number of time steps
dx   = L/(nx-1);
dt   = tmax/(nt-1);
alpha= 3.6*10^-13*24*3600;          % Diffusion constant in m² day⁻¹
Z    = 0;                           % Eroded/deposited layer in meter
% --- Create arrays to save data for export ---%
x = linspace(0,L,nx)'; t = linspace(0,tmax,nt); U = zeros(nx,nt);
% ------------------- Set IC and BC -----------------%
J1    = 0.000034826*24*60*60;       % Bq m⁻² day⁻¹
C_O   = (J1/sqrt(L1*alpha))*exp(-sqrt(L1/alpha).*(x+Z));
U(:,1) = C_O;
% ----Coefficients of the tridiagonal system ---%
a    = (-alpha/(2*dx^2))*ones(nx,1);   % subdiagonal a: coefficients of phi(i-1)
c    = a;                              % superdiagonal c: coefficients of phi(i+1)
b    = ((1/dt)+(alpha/dx^2)+ (L1/2))*ones(nx,1); % diagonal b: coefficients of phi(i)
b1   = -(alpha/dx^2+ L1/2)*ones(nx,1);
c(1) = -alpha/dx^2 ;                   % Fix coefficients of boundary nodes
a(end)= -alpha/dx^2 ;                  % Fix coefficients of boundary nodes
[e,f] = tridiagLU(a,b,c);              % Get LU factorization of coefficient matrix
% ------------------Loop over time steps-----------------%
% ---Right hand side includes time derivative and CN terms ---%
for m=2:nt
d    = U(:,m-1)/dt - [0; a(2:end-1).*U(1:end-2,m-1); 0] ...
     + [0; (b1(2:end-1)).*U(2:end-1,m-1); 0]- [0; c(2:end-1).*U(3:end,m-1); 0];
```

```
d(1)    = U(1,m-1)/dt +( b1(1)*U(1,m-1))+ ((alpha/dx^2)*U(2,m-1))+(2*J/dx);
d(end) = (U(nx,m-1)/dt )- (a(end)*U(nx-1,m-1)) + (b1(end)*U(nx,m-1));


U(:,m) = tridiagLUSolve(d,a,e,f,U(:,m-1));    % solve the system
End
```

% --- Compare with exact solution at the end of the simulation---%

```
t1 = exp(-sqrt(L1/alpha)*x).*erfc((x./(2*sqrt(alpha*tmax)))-sqrt(L1*tmax));
t2 = exp(sqrt(L1/alpha)*x).*erfc((x./(2*sqrt(alpha*tmax)))+ sqrt(L1*tmax));
ue = ((J/(2*sqrt(L1*alpha)))*(t1-t2))+C_O;
plot(U(:,nt),x,'o--',ue,x,'-');
xlabel('Be-7 concentration (Bq m^{-3})');
ylabel('depth(m)');
set(gca,'YDir','reverse','XAxisLocation','top');
```

%-------Calculating Be-7 inventories from the numerical---------%
%--------   simulations   --------%

```
dx1(1:nx)= dx;        %Total number of depth steps
A1     = U(:,nt);     %Corresponding Be-7 inventories
A2     = U(:,nt)'.*dx1;
A      = sum(A2)      %Total Be-7 inventory at the end of the simulation
```
%***%

Appendix G: Uncertainty analysis for diffusion model

The uncertainty propagation which was performed on the diffusion model for steady state and non steady state conditions given by equations (20) and (22) is presented in this section.

For the steady state approach, the measurement uncertainties of the gamma spectroscopy for the activities of ^{7}Be at the reference sites (A_{ref}) and the measurement points (A_s) on the study plot along with the uncertainty on the diffusion coefficient (D) was propagated in the equation (20).

For the non- steady state situations the uncertainties on the simulated inventories were estimated by using the Monte Carlo technique. The uncertainties estimated on the simulated inventories (A_{Num}) were propagated in the equation (22) along with the uncertainties on the measurement points on the study plot and diffusion coefficient (D).

G1. Uncertainties on Diffusion model

The eroded depths estimated under steady state and non-steady state conditions using the diffusion model are given as

$$\Delta z = \sqrt{\frac{D}{\lambda}} \cdot \ln\left(\frac{A_{Ref}}{A_s}\right) \qquad (B42)$$

$$\Delta z = \sqrt{\frac{D}{\lambda}} \cdot \ln\left(\frac{A_{Num}}{A_s}\right) \qquad (B43)$$

The uncertainty on the eroded/deposited depth Δz is calculated by using Gauss error propagation formula given below:

$$\sigma^2_{\Delta z} = \left(\frac{\partial \Delta z}{\partial A_{Ref}}\right)^2 \sigma^2_{Ref} + \left(\frac{\partial \Delta z}{\partial A_s}\right)^2 \sigma^2_s + \left(\frac{\partial \Delta z}{\partial D}\right)^2 \sigma^2_D \text{(Steady state)} \qquad (B44)$$

$$\sigma^2_{\Delta z} = \left(\frac{\partial \Delta z}{\partial A_{Num}}\right)^2 \sigma^2_{Num} + \left(\frac{\partial \Delta z}{\partial A_s}\right)^2 \sigma^2_s + \left(\frac{\partial \Delta z}{\partial D}\right)^2 \sigma^2_D \text{(Non-steady state)} \qquad (B45)$$

Where,

σ_{Ref} = Uncertainty on the ^{7}Be inventory at reference site (Bq m^{-2}) A_{Ref}
σ_{Num} = Uncertainty on the ^{7}Be simulated inventory at a sampling point before erosion (Bq m^{-2}) A_{Num}
σ_D = Uncertainty on the Diffusion coefficient (m^{2}s^{-1})D
σ_S = Uncertainty on the ^{7}Be inventory at a sampling point (Bq m^{-2})A_s

σ_{Ref} and σ_S were obtained from the gamma spectroscopic measurements of the soil samples. σ_{Num} is the uncertainty on the simulated inventory (Discussed in detail in G2). Using (B42) and (B43) in (B44) and (B45) the uncertainties on the erosion/deposition rates are estimated and presented below.

$$\sigma_{\Delta z} = \sqrt{\left(\frac{\sigma_D}{2\sqrt{D\lambda}}\ln\left(\frac{A_{Ref}}{A_s}\right)\right)^2 + \left(\frac{\sigma_{Ref}}{A_{Ref}}\sqrt{\frac{D}{\lambda}}\right)^2 + \left(\sqrt{\frac{D}{\lambda}}\frac{\sigma_S}{A_S}\right)^2} \text{ (steady state)} \quad (B46)$$

$$\sigma_{\Delta z} = \sqrt{\left(\frac{\sigma_D}{2\sqrt{D\lambda}}\ln\left(\frac{A_{Ref}}{A_s}\right)\right)^2 + \left(\frac{\sigma_{Num}}{A_{Num}}\sqrt{\frac{D}{\lambda}}\right)^2 + \left(\sqrt{\frac{D}{\lambda}}\frac{\sigma_S}{A_S}\right)^2} \text{ (non-steady state) (B47)}$$

G2. Uncertainty analysis using a Monte-Carlo method for simulated inventories of ^7Be (σ_{NUM})

A common but very informative Monte Carlo analysis method was used to determine uncertainties on the simulated inventories. First, normal distributions with specific characteristics e.g. mean and standard deviations were assumed for the input variables D and J, in the diffusion model given by equation (5). The diffusion model simulation was then run 200 times with sets of input values randomly selected from these distributions. A cumulative probability distribution of output values is then obtained (Figure 54).

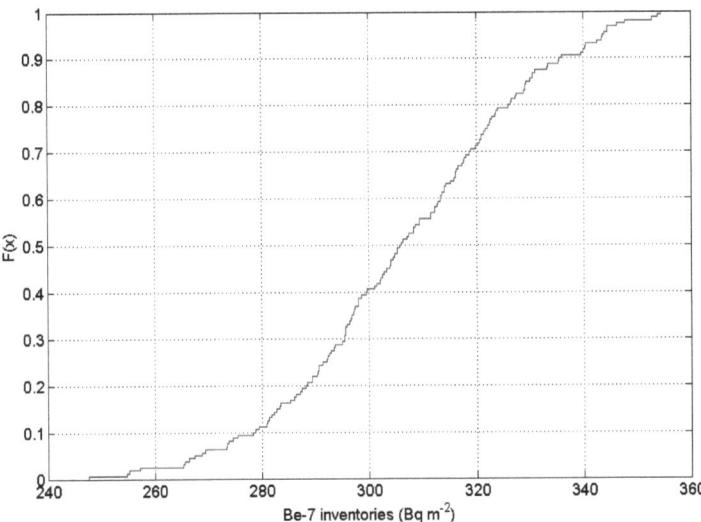

Figure 54: An empirical probability density function for simulated inventories of ^7Be

The input and output normal distributions were standardized using the properties of the distributions such as mean, standard deviation, coefficient of variation CV (= standard deviation/ mean). This allowed the interpretation of the behaviour of error propagation through the modelling process and the contribution of the error of each input value to the overall uncertainty in the model predictive outcome. The confidence intervals of the standard deviations and the means were estimated from the distribution of the model output.

The standard deviations obtained from these simulations were taken as the propagated uncertainties of the simulated inventories of ^{7}Be and were denoted as σ_{NUM}. This procedure was repeated to estimate the uncertainties on simulated inventories of ^{7}Be for all the events and at all measurement points. An example to demonstrate the Monte Carlo approach of uncertainty analysis is presented for the erosion event occurred in the month of Sept-2 in 2011 in Table 1 below.

Table 1 show that the uncertainties estimated by Monte Carlo technique on the total inventories of ^{7}Be have the same order of uncertainty as those propagated from input parameters. The statistical characteristics given in the Table 23 suggest that simulated ^{7}Be inventories for given random distributions produced outputs with CV of 7%.

Table 23: **Summary of statistical characteristics of normal distributions of input and output variables for simulating the inventories of ^{7}Be during the event in Sept-2 in 2011.**

Model inputs	Units	Mean	SD*	CV (%)*	CI*
Diffusion coefficient, D	m^2 day^{-1}	3.0 × 10^{-8}	0.7 × 10^{-8}	23	[0.6 0.8] × 10^{-8}
Atmospheric flux, J	Bq m^{-2} day^{-1}	4.5	0.6	13	[0.5 0.7]

Model outputs	Units	Mean	SD*	CV (%)*	CI
Simulated ^{7}Be inventory	Bq m^{-2}	307	22	7	[20 25]

*SD = standard deviation, CV = coefficient of variability, CI = confidence intervals on the standard deviation for α = 95%.

Appendix H: Statistical fitting technique using non-linear regression

A home made MATLAB program for non-linear regression was used. A function defined by equation (8) estimates the coefficients of a nonlinear regression using least squares. Here, y is a vector of response (dependent variable) values. Typically, X is a design matrix of predictor (independent variable) values, with one row for each value in y. However, X can be any array that function can accept. Provided the initial estimates for coefficients, function returns a vector y of fitted y values.

```
%***************************************************************%

%%******Matlab routine for non linear curve fitting******%%

clc;
clear;
Z = [0.5 1.5 2.5 3.5 4.5 5.5]*10^-3;
C1= [110000 67827 24217 18590 6230 0];
C1e=[15850 18797 6919 7796 3317 0];
C2e=[0.15 0.15 0.15 0.15 0.15 0.15]*10^-3;
L = (0.693/(53.2*24*60*60));
DiffFun = @(p,Z)p(1)/sqrt(p(2)*L)*exp(-sqrt(L/(p(2))))*Z);
startingVals = [0.00001 10^-13];

%*******Confidence intervals for the Parameters*******%

[coefEsts,rw,Jw]= nlinfit(Z,C1,DiffFun,startingVals);
bCIw = nlparci(coefEsts,rw,Jw);
%*******Uncertainties on parameters*********%

    dfe = [size(C1,2)- size(coefEsts,2)];
 [Qw,Rw] = qr(Jw,0);
    msew = sum(abs(rw).^2)/(length(coefEsts));
    Rinvw = inv(Rw);
    Sigmaw = Rinvw*Rinvw'*msew;
    resnorm= sum(rw.^2);
    se   = sqrt(sum(Rinvw.*Rinvw,2)*resnorm/dfe);
%*******Plotting the results*********%

xgrid = linspace(min(Z),max(Z),100)';
[yFitw, deltaw] = nlpredci(DiffFun,xgrid,coefEsts,rw,Jw);
```

```
plot(C1,Z,'ko', yFitw,xgrid,'b-',yFitw+deltaw,xgrid,'b:',yFitw-deltaw,xgrid,'b:');
hold on;
errorbar_x(C1,Z,C1e,'.k')
errorbar(C1,Z,C2e,'.r')
set(gca, 'XAxisLocation', 'top');
set(gca,'Xlim',[0 140]*10^3,'fontsize', 14);
set(gca,'XTick',[0 20 40 60 80 100 120 140]*10^3,'fontsize', 14);
set(gca,'xTickLabel',{'0','20','40','60','80','100','120','140'},'fontsize', 14);
set(gca,'ylim',[0 6]*10^-3,'fontsize', 14);
set(gca,'yTick',[0 1 2 3 4 5 6]*10^-3,'fontsize', 14);
set(gca,'yTickLabel',{'0','0.001','0.002','0.003','0.004','0.005','0.006'},'fontsize', 14);
xlabel('Be-7 concentration (kBq m^-3)');ylabel('Depth (m)');
set(gca,'YDir','reverse');

%%*********Percentages of Be-7 at different depths*********%%
Z1 = [0 1 2 3 4 5 6]*10^-3;
C2 =DiffFun(coefEsts,Z1);
for i= 1:7
    C(i)=C2(i)*100/sum(C2);
end
C3 =cumsum(C);
figure(2)
stairs(C3,Z1);
axes1=gca;
set(gca,'xlim',[0 100],'fontsize', 14);
set(gca,'ylim',[0 6]*10^-3,'fontsize', 14);
set(axes1, 'XAxisLocation', 'top');
xlabel('cumulative distribution of C(%) ');ylabel('Depth (m)'); set(gca,'YDir','reverse');

%******************************************************************%
```

Bibliography

1. Walling, D.E., Quine, T.A. The use of fallout radionuclide measurements in soil erosion investigations. In: Nuclear techniques in soil-plant studies for Sustainable Agriculture and Environment Preservation. Proceedings of International FAO/IAEA Symposium, Vienna. IAEA Proceeding Series STI/PUB/947. Vienna (1995) 597-619.
2. Gallart, F., Lorens, P., Latron, J. Studying the role of old agricultural terraces on runoff generation in a small Mediterranean mountain basins. Journal of Hydrology 159 (1994) 291-303.
3. Gracia-Ruiz, J.M., Lasanta, T., Marti, C., Gonzales, C., White, S., Ortigossa, L. Ruiz, P. Changes in runoff and erosion as a consequence of land-use changes in central Spanish Pyrenees. Physical and Chemical Earth 20 (1995) 301-307.
4. United Nations Environment Program (UNEP). Global Environment Outlook. Earthscan Publications limited (2000) London, UK.
5. Ritchie, J.C., McHenry, J.R. Application of radioactive fallout cesium-137 for measuring soil-erosion and sediment accumulation rates and patterns - a review. Journal of Environmental Quality 19 (1990) 215-233.
6. Walling, D.E., Quine, T.A. Calibration of caesium-137 measurements to provide quantitative erosion rate data. Land Degradation Rehabilitation 2 (1990) 161-175.
7. World Map of the Status of Human-Induced Soil Degradation Global Assessment of Soil Degradation, International Soil Reference and Information Centre, Wagenuigen (1992).
8. IAEA, Use of ^{137}Cs in the study of soil erosion and sedimentation, IAEA-TECDOC-1028 (1998).
9. Higgitt, D.L. Soil erosion and soil problems. Progress in Physical Geography 15 (1991) 91-100.
10. Loughran, R.J. The measurement of soil erosion. Progress in Physical Geography 13 (1989) 216-233.
11. Young, R.A., Onstad C.A. Predicting particle size composition of eroded soil. Trans. American Society of Agricultural Engineers 19 (1976) 1071-1075.
12. Govers, G., Quine, T.A., Desmet, P.J.J., Walling, D.E. The relative contribution of soil tillage and overland flow erosion to soil redistribution on agricultural land. Earth Surface Processes Landforms 21 (1996) 929-946.
13. Meyer, L.D., Foster, G.R., Romkens, M.J.M. Mathematical simulation of upland erosion using fundamental erosion mechanics. Processes of sediment yield workshop, USDA sedimentation laboratory, Oxford 40 (1975) 177-189.
14. United Nations Environment Programme (UNEP). Guidelines for Sediment Control Practices in the Insular Caribbean. United Nations Environment Programme Caribbean Environment Programme (1994) Kingston, Jamaica.
15. Young, R.A., Onstad C.A. Characterization of rill and interrill eroded soil. Trans. American Society of Agricultural Engineers 21 (1978) 1126-1130.

16. Frere, M.H., Woolhiser, J.H., Caro, B.A., Wischmeier, W.H. Control of non point water pollution from agriculture: Some concepts. Journal of Soil Water Conservation 32 (1977) 260-264.
17. Young, R.A., Weirsma, J.W. Relation of rainfall impact on soil detachment and transport. Water Resources Research 9 (1973) 1629-1636.
18. Schuller, P., Iroume, A., Walling, D., Manchilla, B.H., Castillo, A., Trumper, E.R. Use of Berillium-7 to document soil redistribution following forest harvest operations. Journal of Environmental Quality 35 (2006) 1756-1763.
19. Schuller, P., Sepulveda, A., Castillo, A., Walling, D. Use of ^{7}Be to document soil erosion associated with a short period of extreme rainfall. Journal of Environmental Radioactivity 99 (2008) 35-49.
20. Wallbrink, P. J., Murray, A. S. Use of fallout radionuclides as indicators of erosion processes. Hydrological Processes 7 (1993) 297-304.
21. Walling, D.E., He, Q., Blake, W. Use of ^{7}Be and ^{137}Cs measurements to document short- and medium-term rates of water induced soil erosion on agricultural land. Water Resources Research. 35(12) (1999) 3865-3874.
22. Walling, D.E., Woodward, J.C. Use of radiometric fingerprints to derive information on suspended sediment sources. In Erosion and Sediment Transport Monitoring Programmes in River Basins. International Association of Hydrological Sciences 210 (1992) 153-164.
23. Zapata, F. Field application of the Cs-137 technique in soil erosion and sedimentation. Soil and Tillage Research (special issue) 69 (2003)1-153.
24. Zapata, F. Use of environmental radionuclides to monitor soil erosion and sedimentation in the field, landscape and catchment level before, during and after implementation of soil conservation measures. Science Publishers (2007) 301-317.
25. Zapata, P. Handbook of assessment of soil erosion and sedimentation using environmental radionuclides. Kluwer Academic Publishers, London (2002).
26. Schuller, P., Walling, D.E., Sepulveda, A., Trumper, R.E., Rouanet, J.L., Pino, I., Castillo, A. Use of ^{137}Cs measurements to estimate changes in soil erosion rates associated with changes in soil management practices on cultivated land. Applied Radiation Isotopes 60 (2004) 759-766.
27. Walling, D.E., Bradley, S.B. The use of caesium-137 measurements to investigate sediment delivery from cultivated areas in Devon, UK. IAHS 174 (1988) 325-335.
28. Ritchie, J.C., Ritchie, C.A., Bibliography of publications of ^{137}Cs studies related to on and sediment deposition. ww.ars.usda.gov/Main/docs.htm?docid=15237 (2007).
29. Matisoff, G., Bonniwell, E.C., Whiting, P.J. Soil erosion water sources in Ohio watershed using Be-7, Cs-137 and Pb-210. Journal of Environmental Quality 31 (2002) 54-61.
30. Wallbrink, P. J., Murray, A. S. Distribution and variability of ^{7}Be in soils under different surface cover conditions and its potential for describing soil redistribution processes. Water Resources Research 32 (1996) 467-476.

31. Whiting, P.J., Bonniwell, E.C., Matisoff, G., Depth and areal extent of sheet and rill erosion based on radionuclides in soils and suspended sediment. Geology 29 (2001) 1131-1134.
32. Bettoli, M.G., Cantelli, L., Degetto, S., Tositti, L., Tubertini, O., Valcher, S. Preliminary investigations on ^{7}Be as a tracer in study of environmental processes. Journal of Radioanalytical and Nuclear Chemistry 190 (1995) 137-147.
33. Blake, W., Walling, D.E., He, Q. Fallout ^{7}Be as a tracer in soil erosion investigations. Applied Radiation and Isotopes. 51 (1999) 599-605.
34. Olsen, C.R., Larsen I.L., Lowry, P.D., Cutshall, N.H., Todd, J.F., Wong, G.T.F., Casey, W.H. Atmospheric fluxes and march-soil inventories of ^{7}Be and ^{210}Pb. Journal of Geophysical Research D90 (1985) 10487-10495.
35. Wilson, C.G., Matisoff, G., Whiting, P.J. Short-term erosion rates a ^{7}Be inventory balance. Earth surface Processes and Landforms 28 (2003) 967-977.
36. Roth, K. Soil Physics lecture notes, Institute of Soil Science, University of Hohenheim, Germany (1996).
37. Crank, J. The mathematics of diffusion. Oxford Science Publications (1998) 12.
38. Van Genuchten, M.T., Cleary, R.W. Movement of solutes in soil: computer simulated and laboratory results. In: Bolt, G.H. (Ed.) Soil chemistry, Part B. Physiochemical models, Elsevier, Amsterdam (1979) 349.
39. Rodenas, C., Gomez, J., Quindos, L.S., Fernandez, P.L. ^{7}Be concentrations in the air, rain water and soil in Cantabria (Spain). Applied Radiation and Isotopes 48 (1997) 545-548.
40. Rosner, G., Hötzl, H., Winkler, R. Continues wet-only and dry-only deposition measurements of ^{137}Cs and ^{7}Be: an indicator of their origin. Applied Radiation and Isotopes 47 (1996) 1135-1139.
41. Murray, A.S., Olley, J.M., Wallbrink, P.J. Natural radionuclide behaviour in the fluvial environment. Radiation Protection Dosimetry 45 (1992) 285-288.
42. Wallbrink, P. J., Murray, A. S. Fallout ^{7}Be in south eastern Australia. Journal of Environmental Radioactivity 25 (1994) 213-228.
43. Arnold, J.R., Al-Salih, A. Beryllium-7 produced by cosmic rays. Science 121 (1955) 451-453.
44. Cruikshank, A.J., Cowper, G., Grummitt, W.E. Production of ^{7}Be in the atmosphere. Canadian Journal of Chemistry 34 (1956) 214-219.
45. Papastefanou, C. Beryllium-7 Aerosols in Ambient Air. Aerosol and Air Quality Research 9(2) (2009) 187-197.
46. Tilley, D.R., Cheves, C.M., Godwin, J.L., Hale, G.M., Hofmann, H.M., Kelley, J.H., Sheu, C.G., Weller, H.R. Energy levels of light nuclei A = 5, 6, 7. Nuclear Physics A708 (2002) 3-163.
47. Goel, P.S., Narasappaya, N., Prabhakara, C., Rama Thor, Zutshi, P.K. Study of cosmic ray produced short lived P^{32}, P^{33}, Be^7 and S^{35} in tropical latitudes. Tellus 11 (1959) 91-100.
48. Krishnaswami, S., Benninger, L.K., Aller, R.C., Damm, K.L. Atmospherically-derived radionuclides as tracers of sediment mixing and accumulation in near-

shore marine and lake sediments: Evidence from ^{7}Be, ^{210}Pb and ^{239,240}Pu. Earth Planet Scientific Letters 47 (1980) 307-318.
49. Schumann, G., Stoeppler, M. Beryllium-7 in the atmosphere. Journal of Geophysical Research 68 (1963) 3827-3830.
50. Walton, A., Fried, R.E. The deposition of Beryllium-7 and Phosphorus-32 in precipitation at north temperate latitudes. Journal of Geophysical Research 67 (1962) 5335-5340.
51. Beer, J., Blinov, A., Bonani, G., Finkel, R.C., Hofmann, H.J., Lehmann, B. Use of ^{10}Be in polar ice to trace 11-year cycle of solar activity. Nature 347 (1990) 164-166.
52. Lal, D., Malhotra, P.K., Peters, B. On the production of radioisotopes in the atmosphere by cosmic radiation and their application to meteorology. Journal of Atmospheric and Terrestrial Physics 12 (1958) 306-328.
53. Bhandari, N., Rama. Atmospheric circulation from observations of sodium-22 and other short lived natural radioactivities. Journal of Geophysical Research 68 (1963) 1959-1966.
54. Hötzl, H., Rosner, G., Winkler, R. Correlation of ^{7}Be concentrations in surface air and precipitation with the solar cycle. Naturwissenschaften 78 (1991) 215-217.
55. Lal, D., Peters, B. Cosmic ray produced isotopes and their applications to problems in geophysics. Wilson J.G. (Ed.), Progress in Elementary and Cosmic Ray Physics 6, North Holland, Amsterdam (1962) 1-74.
56. Lal, D., Peters, B. Cosmic ray produced radioactivity on earth. Sitte, K. (Ed.), Encyclopaedia of Physics, Springer Verlag, New York (1967) 661-612.
57. Nagai, H., Tada, W., Kobayashi, T. Production rates of ^{7}Be and ^{10}Be in the atmosphere. Nuclear instruments and Methods in Physics Research B 172 (2000) 796-801.
58. Papastefanou, C. and Ioannidou, A. Beryllium-7 and Solar activity. Applied Radiation and Isotopes 61 (2004) 1493-1495.
59. UNSCEAR. Sources and effects of ionising radiation. United Nations Scientific Committee on the Effects of Atomic Radiation. Report to General Assembly, New York (2000).
60. Bondietti, E.A., Hoffmann, F.O., Larsen, I.L. Air to vegetation transfer rates of natural submicron aerosols. Journal of Environmental Radioactivity 1 (1984) 5-27.
61. Bondietti, E.A., Papastefanou, C., Rangarajan, C. A. Aerodynamic size association of natural radioactivity with ambient aerosols. Radon and its decay products: occurrence, properties and health effects, ACS symposium series 331 Hopke, P.K. (Ed.).American Chemical Society, Washington, DC (1987) 377-397.
62. Bonniwell, E.C., Matisoff, G., Whiting, P.J. Determining the times and distances of particle transit in a mountain stream using fallout radionuclides. Geomorphology 27 (1999) 75-92.

63. Winkler, R., Dietl, F., Frank, G., Tschiersch, J. Temporal variation of ^{7}Be and ^{210}Pb size distributions in ambient aerosol. Atmospheric Environment 32 (1998) 983-991.
64. Reiter, E.R. Stratospheric-tropospheric exchange processes. Review of Geophysical Space Physics 13 (1975) 459-474.
65. Bleichrodt, J.F. Mean tropospheric residence time of cosmic-ray-produced ^{7}Be at north temperate latitudes. Journal of Geophysical Research 83 (1978) 3058-3062.
66. Durana, L., Chudy, M., Masarik, J. Investigation of ^{7}Be in Bratislava atmosphere. Journal of Nuclear Chemistry 207 (1996) 345-356.
67. Rama, Honda, M. Natural radioactivity in the atmosphere. Journal of Geophysical Research 66 (1961) 3227-3231.
68. Bhandari, N., Lal, D., Rama. Vertical structure of the troposphere as revealed by radioactive tracer studies. Journal of Geophysical Research 75 (1970) 2974-2980.
69. Feely, H.W., Larsen, R.J., Sanderson, C.G. Factors that cause the seasonal variation of Beryllium-7 concentrations in surface air. Journal of Environmental Radioactivity 9 (1989) 223-249.
70. Duenas, C., Fernandez, M.C., Liger, E., Carretero, J. Gross alpha, gross beta activities and ^{7}Be concentrations in surface air: Analysis of their variations and prediction model. Atmospheric Environment 33 (1999) 3705-3715.
71. Husain, L., Coffey, P.E., Meyers, R.E., Cederwall, R.T. Ozone transport from stratosphere to troposphere. Geophysical Research Letters 4 (1977) 363-365.
72. Dingle, A.N. Stratospheric tapping by intense convective storms: Implications for public health in the United States. Science 148 (1965) 227-229.
73. Baskaran, M., Coleman, C.H., Santschi, P.H. Atmospheric depositional fluxes of ^{7}Be and ^{210}Pb at Galveston and college station, Texas. Journal of Geophysical Research D98 (1993) 20555-20571.
74. Brown, L., Stensland, G.J., Klein, J., Middleton, R. Atmospheric deposition of ^{7}Be and ^{10}Be. Geochimica et Cosmochimica Acta 53 (1989) 135-142.
75. Ioannidou, A., Papastefanou, C. Precipitation scavenging of ^{7}Be and ^{137}Cs radionuclides in air. Journal of Environmental Radioactivity 85 (2006) 121-136.
76. McNeary, D., Baskaran, M. Depositional characteristics of ^{7}Be and ^{210}Pb in southeastern Michigan. Journal of Geophysical Research D108 (2003) (3-1) - (3-15).
77. Ishikawa, Y., Murakami, H., Sekine, T., Yoshihara, K. Precipitation scavenging studies of radionuclides in air using cosmogenic ^{7}Be. Journal of Environmental Radioactivity 26 (1995) 19-36.
78. Duenas, C., Fernandez, M.C., Liger, E., Carretero, J., Liger, E., Canete, S. Atmospheric deposition of ^{7}Be at a coastal Mediterranean station. Journal of Geophysical Research D106 (2002) 34059-34065.
79. Fogh, C.L., Roed, J., Anderson, K.G. Radionuclide resuspension and mixed deposition at different heights. Journal of Environmental Radioactivity 46 (1999) 67-75.

80. Bachhuber, H., Bunzl, K. Background levels of atmospheric deposition to ground and temporal variation of ^{29}I, ^{127}I, ^{137}Cs and ^{7}Be in the rural area of Germany. Journal of Environmental Radioactivity 16 (1992) 77-89.
81. Caillet, S., Arpagaus, P., Monna, F., Dominik, J. Factors controlling 7Be and 210Pb atmospheric deposition as revealed by sampling individual rain events in the region of Geneva, Switzerland. Journal of Environmental Radioactivity 53 (2001) 241-256.
82. Lal, D., Nijampurkar, V.N., Rajagopalan, G., Somayajulu, B.L.K. Annual fallout ^{32}Si, ^{210}Pb, ^{22}Na, ^{35}S and ^{7}Be in rains in India. Proceedings of Indian Academy of Sciences 88A (1979) 29-40.
83. Harvey, M.J., Matthews, K.M. ^{7}Be deposition in high rainfall area of New Zealand. Journal of Atmospheric Chemistry 8 (1989) 299-306.
84. Nijampurkar, V.N., Rao, D.K., Polar fallout radionuclides ^{32}Si, ^{7}Be and ^{210}Pb and past accumulation rate of ice at Indian station, Dakshin Gangotri, East Antarctica. Journal of Environmental Radioactivity 21 (1993) 107-117.
85. Young, J.A., Silker, W.B. Aerosol deposition velocities on the Pacific and Atlantic oceans from ^{7}Be measurements. Earth and Planetary Sciences Letters 50 (1980) 92-104.
86. Vesely, J., Benes, P. Sevcik, K. Occurrence and speciation of beryllium in acidified freshwaters. Water Resources Research 23 (1989) 711-717.
87. You, C.-F., Lee, T., Li, Y.H. The partition of ^{7}Be between soil and water. Chemical Geology 77(1988) 105-118.
88. Olsen, C.R., Larsen I.L., Lowry, P.D., Cutshall, N.H. Geochemistry and deposition of ^{7}Be in river-estuarine and coastal waters. Journal of Geophysical Research 91 (1986) 896-908.
89. Vogler, S., Jung, M., Mangini. Scavenging of ^{234}Th and ^{7}Be in Lake Constance. Limnol Oceanography 41(1996) 1384-1393.
90. Dominik, J., Schuler, C., Santschi, P.H. Residence times of ^{234}Th and ^{7}Be in Lake Geneva. Earth Planet Science Letters 93 (1989) 345-358.
91. Schuler, C., Wieland, E., Santschi, P.H., Sturm, M., Lueck, A., Bollhalder, S., Beer, J., Bonani, G., Hofmann, H.J., Suter, M., Wolfli, W. A multitracer study of radionuclides in lake Zürich, Switzerland. Comparison of atmospheric and sedimentary fluxes of ^{7}Be, ^{10}Be, ^{210}Pb, ^{210}Po, and ^{137}Cs. Journal of Geophysical Research 96 (1989) 17051-17065.
92. Kaste, J.M., Norton, S.A., Fernandez, I.J., Hess, C.T. Delivery of cosmogenic ^{7}Be to forested ecosystems in Maine, USA. Geological Society Abstracts 31 (1999) A305.
93. Huh, C.A., Su, C.C. Distribution of fallout radionuclides (^{7}Be, ^{137}Cs, ^{210}Pb and ^{239,240}Pu) in soils of Taiwan. Journal of Environmental Radioactivity 77 (2004) 87-100.
94. Mabit, L., Benmansour, M., Walling, D.E. Comparative advantages and limitations of the fallout radionuclides ^{137}Cs, ^{210}Pb, and ^{7}Be for assessing soil

erosion and sedimentation. Journal of Environmental Radioactivity 99 (2008) 1799-1807.
95. Kaste, J.M. Dynamics of cosmogenic and bedrock-derived beryllium nuclides in forested ecosystems in Maine, USA. Unpublished M.Sc. Thesis (1999) University of Maine.
96. Owens, P., Walling, D.E. Spatial variability of Cs-137 inventories at reference sites: an example from two contrasting sites in England and Zimbabwe. Application of Radiation Isotopes 47 (1996) 699-707.
97. Pegoyev, A.N., Fridman, S.D. Vertical profile of Cs-137 in soils. Pochvovedeniye 8 (1978) 77-81.
98. Schuller, P., Ellies, A., Kirchner, G. Vertical migration of fallout ^{137}Cs in agricultural soils from southern Chile. Science of the Total Environment 193 (1997) 197-205.
99. Kirchner, G. Applicability of compartmental models for simulating the transport of radionuclides in soil. Journal of Environmental Radioactivity 38 (1998) 339-352.
100. Kirchner, G. Modelling the migration of fallout radionuclides in soil using a transfer function model. Health Physics 74 (1998) 80-85.
101. Bossew, P., Kirchner, G. Modelling the vertical distribution of radionuclides in soil. Part 1: the convection-dispersion equation revisited. Journal of Environmental Radioactivity 73 (2004) 127-150.
102. Ivanov, Y.A., Lewyckyj, N., Levchuk, S.E., Prister, B.S., Firsakova, S.K., Arkhipov, N.P., Arkhipov, A.N., Kruglov, S.V., Alexakhin, R.M., Sandalls, J. Askbrant, S. Migration of ^{137}Cs and ^{90}Sr from Chernobyl fallout in Ukrainian, Belarusian and Russian soils. Journal of Environmental Radioactivity 35 (1997) 1-21.
103. Kirchner, G., Baumgartner, D., Delitzsch, V. Schnabl, G., Wellner, R. Laboratory studies on sorption behavior of fallout radionuclides in agricultural used soils. Modelling Geo-Biosphere Processes 2 (1998) 115.
104. Deumlich, D., Barkusky, D. Sudangras - eine Alternative zum Energie-Mais. Workshop der Kommissionen VI und IV der DBG+DGP (2010).
105. http://www.zalf.de/home_zalf/institute/zentral/fs/fs/fsm/index.html
106. Schkade, U. World-wide open proficiency test for ^{226}Ra in soil. IAEA (2009).
107. Schkade, U. World-wide open proficiency test for radioactive samples in the environment. IAEA (2010).
108. http://www.bmu.de/strahlenschutz/ueberwachung_der_umweltradioaktivitaet/m essanleitungen/doc/41754.php/files/pdfs/allgemein/application/pdf/strlsch_mes sungen_k08.pdf
109. Yang, M. Y., Walling, D.E., Tian, J.L., Liu, P.L. Partitioning the contributions of sheet and Rill erosion using Beryllium-7 and Cesium-137. Soil Science Society of America Journal 70 (2006) 1579-1590.
110. Bergsma, E. Terminology for soil erosion and conservation. International Society of Soil Science (2000).

111. Ratzke, U. Beiträge zum Bodenschutz in Mecklenburg-Vorpommern. Bodenerosion. Landesamt für Naturschutz und Geologie (2002).
112. Bodenerosion durch Wasser- Kartieranleitung zur Erfassung aktueller Erosionsformen. DVWK MEKRBLÄTTER (1996) 239.
113. Mabit, L., Toloza, A., Nirschl, A. Development of a Fine Soil Increment Collector (FSIC) to solve the main limitation of the use of ^{7}Be as soil tracer. IAEA-Soils Newsletter, Laboratory Activities-Research, 30 (2008) 21-22.
114. Carslaw, H.S., Jaeger, J.C. Conduction of heat in solids. Oxford Science Publications (1997) 75.
115. Recktenwald, G.W. Finite-Difference Approximations to the Heat Equation. Mechanical Engineering Department Portland State University, Portland Oregon (2004).
116. http://www.hydroskript.de/html/ index.html?page=/html/hykp0404.html
117. Toy, T., Foster, G. and Renard, K. Soil Erosion: Processes, Prediction, Measurement, and Control. John Wiley & Sons, Inc. (2002).

I want morebooks!

Buy your books fast and straightforward online - at one of the world's fastest growing online book stores! Environmentally sound due to Print-on-Demand technologies.

Buy your books online at
www.get-morebooks.com

Kaufen Sie Ihre Bücher schnell und unkompliziert online – auf einer der am schnellsten wachsenden Buchhandelsplattformen weltweit! Dank Print-On-Demand umwelt- und ressourcenschonend produziert.

Bücher schneller online kaufen
www.morebooks.de

VDM Verlagsservicegesellschaft mbH
Heinrich-Böcking-Str. 6-8
D - 66121 Saarbrücken

Telefax: +49 681 93 81 567-9

info@vdm-vsg.de
www.vdm-vsg.de

Printed by Books on Demand GmbH, Norderstedt / Germany